A Force of Nature

General Editors: Edwin Barber and Jesse Cohen

President Reagan: The Triumph of Imagination

President Nixon: Alone in the White House

President Kennedy: Profile of Power

What the People Know: Freedom and the Press

Passage to Peshawar

American Journey: Travelling with Tocqueville in Search of Democracy in America

Convention

A Ford, Not a Lincoln

A Force of Nature: The Frontier Genius of Ernest Rutherford

GREAT DISCOVERIES

RICHARD REEVES

A Force of Nature

The Frontier Genius of Ernest Rutherford

ATLAS & CO.

W. W. NORTON & COMPANY

NEW YORK · LONDON

First published as a Norton paperback 2008

For information about permission to reproduce selections
from this book, write to Permissions,
W. W. Norton & Company, Inc., 500 Fifth Avenue, New York, NY 10110

For information about special discounts for bulk purchases, please contact
W. W. Norton Special Sales at specialsales@wwnorton.com or 800-233-4830

Book design by Chris Welch
Production manager: Julia Druskin

Library of Congress Cataloging-in-Publication Data

Reeves, Richard, 1936–
A force of nature : the frontier genius of Ernest Rutherford /
Richard Reeves. — 1st ed.
p. cm.
Includes bibliographical references and index.
ISBN 978-0-393-05750-8 (hardcover)
1. Rutherford, Ernest, 1871–1937. 2. Physicists—England—Biography.
3. Physicists—New Zealand—Biography. 4. Science—England—History—
20th century. 5. Science—New Zealand—History—20th century.
6. Nuclear fission. 7. Radioactivity.
I. Title.
QC16.R8R44 2008
530.092—dc22
[B]

2007033184

ISBN 978-0-393-33369-5 pbk.

Atlas & Co.
15 West 26th Street, New York, N.Y. 10010

W. W. Norton & Company, Inc.
500 Fifth Avenue, New York, N.Y. 10110
www.wwnorton.com

W. W. Norton & Company Ltd.
Castle House, 75/76 Wells Street, London W1T 3QT

1 2 3 4 5 6 7 8 9 0

*This book is for Ian, Alex and Aidan Fyfe,
and Rory Catherine O'Neill.*

A Force of
Nature

Introduction

On a seventh-floor hallway wall of the physics building at Stevens Institute of Technology, one of the oldest engineering schools in the United States, there is a poster published by the American Physical Society titled "A Century of Physics." It reads, in part:

> By the end of the 19th Century, after more than 2000 years of struggle that began with the Greek philosophers, physical scientists had reason to believe that they were beginning to understand the Universe. Their theories of matter and energy, of electricity and magnetism, of heat and sound and light were confirmed in laboratories throughout the world with increasing precision. Experimentation was the method, and mathematics the language of a powerful, coherent body of knowledge called classical physics.
>
> In 1911, the atomic nucleus was found. To explain the recoil of alpha particles from thin gold foils, New Zealand born physicist Ernest Rutherford, working in England, proposed the nuclear model of the atom.

Ernest Rutherford, a big, boisterous country boy from the frontier of New Zealand, created the science we call nuclear physics. He made a new world, different from what the Greeks believed, different from what had been suggested by Sir Isaac Newton, the greatest of classical physicists. Rutherford, born in 1871, fifteen mountain miles from the nearest town, proved that the atom was not the unseeable, indivisible solid sphere that made up all the world and universe. Working with his hands, the simplest of mechanical equipment, and, most of all, a mind plowing ahead like a battleship, the young New Zealander found and explored a new world—a subatomic world of swirling electric particles and forces beyond most human comprehension. A charismatic man not above self-promotion, Rutherford later said, "I have broken the machine and touched the ghost of matter."[1]

I had graduated from Stevens, with a degree in mechanical engineering. A thousand times and more I had walked up the hill to Castle Point, a stunning promontory overlooking the towers of mid-Manhattan across the Hudson River. I had stood on that point with hundreds of my classmates on a clear night in 1957 watching a Soviet satellite, Sputnik, a dart of light moving across the darkness. We and much of the world were filled with the dread that the forces revealed by Rutherford and physicists who followed him—one of his students, a Russian named Pyotr Kapitsa, was said to be the principal designer of the Soviet vehicle[2]—could now be launched from space and destroy the planet itself. I had also traveled with those same classmates to new nuclear energy plants, listening to their designers urging us to help them harness those same atomic forces for the good of all humanity.

I returned to the school more than forty years later to work with the Physics Department to re-create the experiment that Rutherford had first used to "see" into the atom and then map out or imagine the structure we know—a tiny universe, a vacuum, with electrons orbiting a highly charged, incredibly dense nucleus so small that it was said by Rutherford to be the equivalent to a pinhead in the earthly vastness of St. Paul's Cathedral. So, on November 5, 2005, I was sitting in the dark in a laboratory, Room 619, with the four other men who, over many months, had designed from scratch the apparatus that made the re-creation possible. We were in the dark, as Rutherford and two assistants had been almost one hundred years before, to adjust our eyes before looking through a microscope into a vacuum chamber hoping to see sparks of light, radioactive alpha rays striking a phosphorescent wall after passing through a bit of gold foil. 3

This "scattering experiment," as it is known, is what scientists call an elegant piece of work, which, in less sophisticated conversation, means that it can be described rather simply. Radioactive particles are fired through the foil at the phosphorescent target, a plate painted with zinc sulfide. "Scintillations" can be seen through the microscope: particles hitting the plate. "Alpha particles," Rutherford decided to call them. According to classic theory, these particles should have sparked along a straight line mirroring a small slit in the lead box encasing the radioactive source. In early experiments, Rutherford had noticed that the edges on the target were sometimes fuzzy. That fuzziness should not have been possible if the atom was what scientists then thought it was, a kind of ball of mixed positive and negative electric charges—nothing powerful enough or dense enough to deflect particles. But something in the gold atoms that made up the foil was doing just that: deflecting the alpha projectiles just a bit. Or maybe more than a bit. The microscope was focused on the line. Was it possible that some of the alpha particles were being deflected at wider angles, outside the target that Rutherford could see?

3 Rutherford told his two young assistants, Hans Geiger and Ernest Marsden, to reconfigure the apparatus to determine whether any particles were hitting the plate farther from the line. They were. A few of the particles scattered at angles as great as 90 degrees and more. About one in eight thousand particles actually bounced back toward the source.

3 Rutherford, then a professor at the University of Manchester in England, went on to other things. He was a young man, thirty-eight, and a busy man—famous already, a Nobel Prize winner, and head of one of the world's great laboratories. But he continued to think, hard and slowly, about the scattering. Then, at lunch on a Sunday two years later, he looked up and said, "I know what the atom looks like."[3] This book is about that revelation, and the man who made it all happen.

And that is why the five of us were sitting in darkness: I was there with Professor Kurt Becker of the Physics Department, as well as two talented graduate students (Frank Corvino and Damien Marianucci) and the shop superintendent (George Wohlrab), who sounded almost exactly like Rutherford's shop stewards—skilled mechanics amused and frustrated by the ways of professors.

Rutherford's men said that their conversations in the dark with "The Prof," as he was sometimes called, were the most interesting of their lives. Our talk on this day in 2005 may not have been as stimulating, but we certainly had a good time telling each other stories about science, professors past and

present, travel, food—American research is driven by pizza—
and the many problems we encountered as we tried to duplicate what Rutherford, Geiger, and Marsden had done in 1909.
"Time travel," said Corvino.

"I remember our very first conversation," said Becker. "The
starting point was to get as close to the original as possible.
But as we moved along, we quickly realized that there were a
lot of things that they were able to get away with that you can
no longer do today because of safety concerns."

In 2005, we had a couple of things going for us. First, of
course, we knew how the experiment was supposed to come
out. They had been out there in the blue. We also had faster,
stronger, and more reliable equipment to pull vacuums,
greatly reducing the chances of results contaminated by the
odd air molecule or hydrogen atom. But we, and the world,
now knew the deadly dangers of radioactivity. Rutherford
used to toss bits of radioactive material in his pocket and
then, before dinner, into the top drawer of a desk at home. We
were dealing with strict international regulations and government and school inspectors protecting us against our own
curiosity and enthusiasm. In the end the source of our projectiles (the alpha particles) was a distant descendant of radium:
americium-241. Its radioactivity measured 25 millicuries,*

---

* Radioactivity is measured by the "curie"—named for Marie Curie, the
discoverer of radium—which represents 23.7×10^{10} disintegrations per second. Americium, an artificial element with an atomic number of 241, was
produced in 1944 by the American scientists Glenn Seaborg, Ralph James,
Leon Morgan, and Albert Ghioso at the University of Chicago. It is a product of the radioactive decay of plutonium-241, which is about 12 percent of
the 1 percent content of plutonium in typical spent fuel from a nuclear
power reactor. Plutonium has a half-life of 14 years. One of the decay prod-

significantly weaker than Rutherford's original source: radon-222, a radioactive gas produced by the decay of thorium, a metal just a few steps down the evolutionary radioactive ladder from radium. One indication of the potency of the radon used in 1908 was that it produced 30 billion alpha particles per second. Our americium was producing perhaps 370 million particles per second. "If we tried with what they used, we couldn't all be in this room," Becker said. "We couldn't be in the building."

George Wohlrab and Damien Marianucci, who did most of the building of the apparatus, did what builders love to do: talk about the problems they had overcome, and how professors never understand or appreciate what they do. "The nine-finger professors," someone said, talking about a man we all knew who had gone into Wohlrab's shop alone one weekend and cut off the tip of a pinkie while trying to machine a small part critical to his own experiments.

"It was a piece of crap when you gave it to me," the machine shop foreman said of the vacuum chamber we were using, a beautiful stainless steel cylinder originally machined at Queens University in Belfast, Northern Ireland. Becker had brought it with him as he climbed the academic ladder from Belfast to Windsor, Canada, Lehigh University in Pennsylvania, and the City University of New York before crossing the river to Stevens, where he had become head of the Physics Department.

---

ucts over those years is americium-241, a weaker source that has its own half-life of 432 years and that, during those years, is emitting alpha particles to become neptunium-237. Americium-241, in almost microscopic amounts, is in your house, the working ingredient of common commercial smoke detectors.

Wohlrab, as is his habit and heritage, was mocking the mechanical skills of the rest of us, complaining that he had had to grind and sand away the scratches of travel and use. Scratches on the flanges are leaks; they break the vacuum. "It's amazing the way they could do it back then," Wohlrab said, "using glass instead of metal. Blowing the glass, grinding it."

"I learned to blow glass here," I said. I could feel heads turning toward me in the dark. "I learned too. We had to in Germany," Becker said. It was probably good that we could not see the faces of Corvino and Marianucci, both in their twenties.

"Welding. Making our own gears," I continued. "We did all that. I learned to weld and cut in the same basement Alexander Calder did. 'Course he made better use of it." The artist had graduated from Stevens in 1900, sixty years before me.

Marianucci said that the problems, as always, continued right up to the end. There had been a leak the day before through the bolt holes on two of the plates capping flanges that we did not need. He said he thought we might have to call off the experiment, and he didn't know whether he was more afraid to tell me or to tell George.

"George," I said.

Wohlrab confirmed that, saying that working with physicists was like trying to build a foundation when all you have is a roof. At any rate, the two graduate students worked out the problems by themselves.

Did they talk this way in the dark all those years ago in a basement at the University of Manchester? Probably. Wohlrab was playing the same role as a man named Jost did in a story told by A. S. Eve, one of Rutherford's researchers: "I was asked by Rutherford to make a sensitive small scale electroscope, the gold leaf of which would remain charged for two or three days. This I failed to do. So Rutherford said: 'Lester Cooke used to make them; why can't you? Get Jost, the mechanic, to make you one.' So I went to Jost and repeated this. He said: 'If I could not make a better electroscope than Cooke, I'd shoot myself.'"[4]

They must have shown more deference to Rutherford, who had won a Nobel Prize for work done not long after his thirtieth birthday. But the subjects were probably the same. Rutherford, whose voice could rattle beakers, would have cut off the conversation as he always did: "Let's get results. I want results!" [3]

"Let's go," said Frank Corvino, who was actually going to run the experiment. "Let's see God." We had been in the dark for more than a half hour. Corvino was in charge now, and his notes read,

> As the 35 minute mark approached, I began to grow nervous. Nervous not because the past year of work might have an anti-climactic ending, but because I knew what was necessary to complete the job. The apparatus was designed with a mirror at a 45-degree angle to the window, and a telescope pointing at the mirror, perpendicular to the window, out of the line of fire of gamma rays emerging from the Americium. After calculating 50–100 scintillations per minute, the chances were remote that a particle would not get absorbed by the optics—the window, the mirror or the lenses. As we looked through the telescope this was confirmed. We saw nothing.
>
> After a short coffee break we knew and discussed what had to be done. The optics had to be removed and the

viewing eye had to look right through the microscope into the window flange, looking directly toward the source. This was the only way, and it seemed fitting because this was the way it was done in 1909. During this dilation period in the dark the conversations were quite different. Instead of past adventures, we discussed how much damage could be done to our eyes and how incredibly strong the Thorium was that Geiger and Marsden used. I couldn't help thinking that ignorance really is bliss. Those two had no idea what the dangers of high frequency waves were. They were just following orders. And in a way so was I. The radiation that we were being exposed to was less than a medical X-ray. But how long would it take to see a scintillation? The exposure time was guaranteed to be longer.

Once our eyes were adjusted, Reeves said: "At my age, how much can this really hurt me." I looked into the window for about 15 seconds before I saw anything and then a flash. Or was I just seeing things? But then another and another. All relatively small, sharp flashes that disappear as fast as they appear. All came around the same area, the center of which is the zero-degree point. In other words, these alpha particles were tunneled directly through the empty space of the gold atoms. It was a great feeling, which was further intensified once Reeves and Damian saw the same.[5]

"A good experiment, " Corvino said later. "We didn't create 100 percent, but re-creating 90 percent is really something. Good experiment." He used the word "fun" several times during the day. We all did. Rutherford used to repeat, "Well, it's a great life!"[6]

It was an extraordinary life. Born in a rain forest at the bottom of the Earth, Ernest Rutherford was, simply, a genius; that is, he changed the way the rest of us see the world and ourselves. He was the first man to show that elements are not immutable; they can naturally transform themselves into other elements—the process we call both "radioactive decay" and "half-life." He discovered the nuclear structure of the atom, beginning what came to be known as the "heroic age" of physics. And he "split the atom." In 1932, he and his "boys" became the first to do that—or, more accurately, to break open the nucleus of the atom and reveal and release forces unimagined, unimaginable. Eleven of his boys—students and researchers—went on to win Nobel Prizes. Asked by newspapermen how they could do that—what was different about them—Rutherford said, "We're like children who always want to take apart watches to see how they work."[7]

The boy from a frontier settlement then called Spring Grove wanted to know what the world was made of and how it worked. This is his story.

CHAPTER 1

On the first day of May 1851 in London's Hyde Park,
Prince Albert, the consort of Queen Victoria, opened
"The Great Exhibition" in a giant glass and steel
building called "The Crystal Palace." It was, in the fashion of
the time, a World's Fair of industry and culture, in that order.
In announcing his sponsorship of the big show as construc-
tion began, the prince said,

> We are living at a period of most wonderful transition
> which tends rapidly to the accomplishment of that great
> end to which, indeed, all history points—the realization of
> the unity of mankind . . . The distances which separated the
> different nations and parts of the globe are gradually van-
> ishing before the achievements of modern invention, and
> we can traverse them with incredible ease; the languages of
> all nations are known and their acquirements placed
> within the reach of everybody; thought is communicated
> with the rapidity and even by the power of lightning . . .

The knowledge acquired becomes at once the property of all of the community at large . . . no sooner is a discovery or invention made, than it is already improved upon and surpassed by competing efforts: the products of all quarters of the globe are placed at our disposal, and we have only to choose what is cheapest and best for our purposes . . . Science discovers these laws of power, motion and transformation; industry applies them to raw matter which the earth yields us in abundance, but which becomes valuable only by knowledge.[1]

That was the idea, and it was a huge success. Fireworks and fountains, one of them shooting water 250 feet into the air, helped attract more than six million visitors who wandered among thirteen thousand exhibits, from giant looms and other industrial marvels from around the world to smaller new things called "kitchen appliances." After six months, exhibit profits totaled more than 186,000 pounds sterling. The money was used to build Albert Hall, the national Science Museum, and the Victoria and Albert Museum. The royal commission that presided over the creation of the exhibit also set aside a small amount to provide annual "Exhibition of 1851 scholarships" to bring talented science students from all over the British Empire to study and do research in the home country.

In 1895, there were two candidates from New Zealand. J. C. Maclaurin was a thirty-one-year-old chemist working on gold extraction processes. The other, seven years younger, was Ernest Rutherford, working on wireless experiments, sending magnetic waves through a big but relatively primitive laboratory at Canterbury College in Christchurch. The commission chose Maclaurin. But Maclaurin decided to marry, which made him ineligible for the prize. Rutherford was picking potatoes in the family garden at Pungarehu, a remote settlement where his father had set up a flax mill, when a mailman came by with the news that the scholarship was his now. Throwing his shovel in the air, the young man shouted, "That's the last potato I'll ever dig."[2]

Ernest Rutherford was born on August 30, 1871, in a house that his father had built in a small settlement on the New Zealand frontier. There were perhaps 250,000 white people and a much smaller number of Maori people, South Pacific natives, sharing the two islands twelve thousand miles from England—the place the whites called "home." James Rutherford, Ernest's father, had come to the Antipodes* in 1843, when he was just five years old, the son of a wheelwright named George Rutherford from Dundee in Scotland. The trip to a new land on a 125-foot sailboat named the *Phoebe Dunbar*, carrying thirty crew members and two hundred immigrants, took six and a half months. That's how far away New Zealand was then. The little wooden houses and small towns made the place look like a verdant version of the American West at the same time.

The Rutherfords were poor, and so was New Zealand. But George and then his son James were known for being good with their hands and plain tools. As boy and man, James used only wood to make the bicycles that he used to get around—a

---

* "Antipode," from the Greek, means "the exact opposite." It is often used in England to describe Australia and New Zealand because they are almost exactly on the opposite side of the Earth from England. The settlement was called Spring Grove at the time but was soon renamed "Brightwater," and that was the name Rutherford always used.

skill that he passed on to his own sons. He was a "bush engineer," able to make almost anything, but those were hard times and he failed at more than one business. Ernest's mother, Martha Thompson Rutherford, was a schoolteacher, an educated woman who owned the symbol of frontier gentility, a Broadwood piano from England that she polished every day. The fourth of the couple's twelve children, Ernest was born in the settlement of Brightwater, south of Nelson on New Zealand's South Island. His father was farming there and trying to make a go of building a flax mill; he had learned the trick of stripping local reeds and leaves from the Maori. He also won contracts to make forty thousand birch ties for new railroads at two shillings, eight pence each. He might have become a rich man, prosperous at least, but depression hit the colony and somehow the railroads were never built.

Through it all, Martha Rutherford, a determined and rigid woman, schooled her children—"All knowledge is power," she told them most days—and sent them, as boarders, to the schools in the settlements of Foxhill and Havelock. Ernest was a big ruddy kid, loud and rowdy, playing a rough forward on school rugby teams. But he also read every book he could get his hands on and studied everything around him, from the stars and plant life to the ways of Maori, who were still a Stone Age people when Captain Cook first saw the Antipodes in 1769. The native South Pacific islanders, like their Polynesian cousins as far north as Hawaii, lived on fish along the coast, mostly on North Island. Astonishingly, the lush and warm islands had birds but no native land animals at all, and no major food crops before the British came with their trading companies, religious groups, and fortune seekers. The

British brought seeds for grain by ship from other parts of the world.

After her son became famous in the early 1900s, Martha Rutherford, along with two of Ernest's sisters, agreed to talk about his childhood for the archives of the University of New Zealand. This is part of what they said:

At one time he had a passion for taking photographs of his family, brothers and sisters, with a home-made camera, and he was continually taking clocks to pieces. Reading was his great recreation. Anything seemed to interest him, even the very lightest of literature. He was especially fond of Dickens when young and would welcome any opportunity to read out loud to younger members of the family, joining most heartily with his huge laughter at the various humourous incidents of "Pickwick." In holiday time, he was often asked to teach his younger sisters . . . In order to hold their attention or keep them quiet, he used to tie their pig-tails together. When asked by his mother to teach the girls something, he would generally reply, "If they are mustered out by 9 o'clock, I will, but you must do the mustering." Saturdays were often employed in bird-nesting, spearing eels in the river, catching brook trout in the bush pools, and long walks . . . Ernest tried his hand at pheasants and wild pigeons. He used to get a horse and ride through the bush, often through deep mud, to reach the objective just as the sun was rising. There were three miro trees, the berries of which attracted the pigeons. They came in large numbers, but the ammunition was home-made and the guns poor, so that he could not bring any down out of the high trees. The birds sitting on the trees formed a very small target, until Ernest suggested firing when the birds were about to alight, which they would do with their wings outspread. On one occasion 16 were bagged.[3]

The smart kid from Brightwater became the most famous of New Zealanders and never lost his love of the place, so stories about him, many true, are still told, in the manner of George Washington in that other British colony along the Atlantic Ocean. In one standard story, Ernest's father wakes in the middle of a stormy night to see his son outside watching a thunderstorm:

> "What's up my boy?" called the father.
> "I'm counting," young Ernest answered. "You can tell how close you are to the center of the storm."
> "Counting?"
> "Yes. You know where the center is if you count the seconds between the flash and the thunder clap. Then you allow 1,200 feet for each second for the sound to travel."[4]

True or not, the story typifies the kind of methodology that Rutherford would use to become a great man of science, the preeminent experimental physicist of the early twentieth century, perhaps of the entire century. He read his first science book at the age of ten, and his mother saved it. The small volume, *Primer of Physics* by Balfour Stewart, professor of physics at Manchester University, began, "This book has been written, not so much to give information, as to endeavour to discipline the mind by bringing it into immediate contact with Nature herself, for which purpose a series of simple experiments are described leading up to the chief truths of each science, so that the powers of observation in the pupils may be awakened and strengthened."[5]

Ernest Rutherford was the best student they ever had at Havelock, and a small crowd gathered when he took the Marlborough Provincial Scholarship examination for Nelson College—high school in the U.S. system—in the small building that was the Havelock school. He rode over the mountains to Nelson on horseback with his father. It was the farthest from home that he had ever been. Friends and distant neighbors stood in the sun that summer day and watched through the windows, waiting for the results, and cheered when it was announced that the local boy had scored 580 out of 600, the highest total ever recorded. He finished first in every subject while serving as "head boy" of the eighty-boy school and knocking heads on the rugby team. Winning new scholarships, he went on to Canterbury College, a school with just seven professors, in Christchurch, the largest city on South Island, which, then and now, looked like a small, provincial English city. He completed a bachelor of arts degree in 1892, spending holidays tutoring his sisters, and then stayed on at Canterbury to earn master's degrees in mathematics and science, followed by a bachelor of science degree with firsts in chemistry and geology. He continued to play rugby and joined the debating club, doing everything except dance, a skill he never mastered.

Then, during one of his trips home, his younger brothers Herbert and Charles went sailing on Pelorus Sound and never returned. For a year, James Rutherford, sometimes accompanied by Ernest, walked the edge of the sound searching for the bodies but never finding them. Ernest's mother never played the Broadwood again. James moved the family to Pungarehu on North Island, setting up a flax mill near the swamps where the plants that he needed grew wild.

In 1888, when James Rutherford moved his family, the swamps of Pungarehu were a long way from Christchurch, and Ernest began living in a boardinghouse in the little city of sixteen thousand people. He started spending time with Mary Newton, his landlady's daughter. Mary was a formidable woman—small like Ernest's mother, well-read, a piano player. Mary's mother was formidable too; she became the head of the local temperance society after her husband died of drink.

While earning money teaching high school classes, Ernest asked Mary to marry him. "That would be idiotic," she said. "You have to finish your education—and decide what you want to do next. I would be a handicap."[6] He asked whether she would marry him later. "Of course I will. I wouldn't dream of marrying anyone else," she replied. That was it. The engagement, secret at first, would last six years.

Ernest spent most of his time in Christchurch conducting electrical experiments, developing a wireless mechanism to capture the electromagnetic waves—called "radio waves" now—discovered in Germany in 1888 by Heinrich Hertz. Reading journals, he was able to re-create Hertz's experiments and results, and then go further. Using batteries that he fashioned himself with zinc electrodes dipped in acid, which had to be made fresh each day, young Rutherford was able to send signals sixty feet through stone walls and an iron door—much to the delight of other Canterbury students, who crowded in to watch the show. His results were published in 1894 in *Transactions of the New Zealand Institute*, which included three specific sections—"Zoology," "Botany," and

"Geology"—providing a fair indication of the state of science then and there. Rutherford's papers were published under "Miscellaneous." 3

However they were classified, the electromagnetic experiments were Rutherford's initiation into the way modern science works best: open distribution and publication of theories and the results of experiments. The results and new information drove new thinking and research by scientists, known and unknown, in places as far away from Cambridge and Berlin as Christchurch.

Young Rutherford's radio experiments were a great and amazing entertainment at Canterbury. After a while, town residents joined students in packed lecture halls to watch Rutherford at work. He was much admired when he taught for a year at Boys High School in Christchurch, but not for his classroom manner. In the university archives, one of his students, a Mr. O. Gillespie, wrote,

He was entirely hopeless as a school master. Disorder prevailed in his classes . . . I do not remember myself following any of his intellectual processes on the blackboard. They were done like lightning. When he detected some more than usually noisy boy he sent him sternly for the Appearing Book. All the lad had to do to escape the consequences of misdeeds was to stay out of the class room long enough for Rutherford's enormous mind to have bulged in some other direction, sneak back to his seat and he would inevitably not be noticed . . . We certainly had him added up as a genial person whose interests were nothing to do with the keeping in order of small boys.[7]

Luckily for all concerned, the Exhibition of 1851 scholarship took Rutherford away from Christchurch in August 1895. He borrowed money, most of it from his older brother, George, a successful farmer, and, carrying a box with his crude radio wave apparatus—still on display today at Cambridge University—boarded ship for the long passage to London. The ship made a stop in Australia, and Rutherford sought out the most famous physicist in the Antipodes: William Henry Bragg of Adelaide University, a man who would play an important part in Rutherford's later life. Three weeks later Rutherford arrived in London, alone and practically broke in the big city. He had the 1851 prize, which provided him with 150 pounds sterling to pay tuition and all living costs, and a few letters of introduction—one from Bragg and others from New Zealand dignitaries.

London was not a good fit. In the soot and damp cold of the big city, Rutherford immediately came down with a bad cold and neuralgia, which kept him in bed at a boardinghouse for several days. Then, in a cruel slap of fate, he slipped on a banana peel, wrenching a knee—an injury that pained him for the rest of his life.

He had, however, sent copies of his published work in New Zealand to the Cavendish Laboratory at Cambridge University. Late in September he received a letter from J. J. Thomson, the English scientist whose book *Notes on Recent Researches in Electricity and Magnetism*, published in 1883, had won him enough acclaim at the age of twenty-seven that he was named director of the laboratory the next year. Thomson's letter began, "I shall be very glad for you to work at the Cavendish, and will give you all the assistance I can . . . If you could spare the time to come to Cambridge, I should be glad to talk matters over with you."[8] Thomson kept that promise and more over a lifetime—friends thought he was touched that a promising young man had come twelve thousand miles to work with him—and Mrs. Thomson became like a second mother to that young man.

Glad to get out of the city, Rutherford took the train to Cambridge, an hour and a quarter away. "The country is pretty enough but rather monotonous," he wrote to his mother. "I went to the Lab. and saw Thomson and had a good long talk with him. He is very pleasant in conversation and is not fossilized at all."[9]

Rutherford's fellow students were quite a different story. The young elite of Cambridge were openly hostile toward Rutherford, as well as toward a young Irishman, J. S. Townsend. Under a new policy, the two young colonials were the first research students admitted to graduate studies at Cambridge who were not graduates of that venerable and enormously self-satisfied institution. The common attitude was that somehow these provincials had cheated or jumped above themselves into the place. Rutherford's provincial work ethic was also something of an affront to the young gentlemen used to the more leisurely pace of the privileged. "In your last letter you mentioned that I hadn't mentioned if I was doing any research work," Ernest wrote to Mary Newton in December 1895. "My dear girl, I keep going steadily turning up to the Lab. five nights out of seven, till 11 or 12 o'clock."[10]

"They snigger at us," said Rutherford of his fellows as they passed. "I'd like to do a Maori war-dance on the chest of one," he wrote to Mary, "and will do that in the future if things don't mend." But after a couple of months it was obvious that the old boys were having trouble keeping up with the new ones, particularly Rutherford. In early December, little more than two months after his arrival, Rutherford saw a notice on Cavendish's main bulletin board announcing a lecture: "A method of measuring along wires and determination of their period . . . with experiments by E. Rutherford."

It was quite an honor for a new boy, and apparently it went well, with Rutherford repeating some of his New Zealand experiments. "My friends all reckoned I did very well indeed," he wrote to Mary. "Mrs. J. J. in speaking to me afterwards complimented me rather neatly in a way (which of course I took *cum grano salis*) that struck me as decidedly good, 'You kept us ladies very interested indeed, and I am sure it was sufficiently deep for more scientific members of the society.'"[11]

In fact, it was too deep for some. Rutherford continued in his letter,

> The three demonstrators* are extremely friendly now they see we have made a strong position for ourselves and I grimly rejoice for they did not take any notice of us the first two months . . . My paper before the Physical Society was a heavy blow to their assumed superiority and now they all offer to help us in anyway they can and tell me confidentially about their own little researches—so wags the world . . . I must apologize for the amount of ego that fills this letter, but human nature will out."[12]

One of the young men who worked with or, at least beside,

---

* The "demonstrators" would be called "instructors" or "graduate assistants" in American universities. They were paid the sterling equivalent of between 500 and 750 dollars a year.

Rutherford—Henry Dale, who would become an important British scientist himself—later wrote,

> We ordinary Cambridge students, like a lot of our seniors, were inclined to look a little askance at these representatives of a new species . . . Rutherford was probably the most brilliant of them all though we might not have recognized him as such for ourselves . . . I think that we young cynics might have described his manner as rather hearty and even a little boisterous, but I can imagine us allowing, with pride in our tolerance, that a man who had been reared on a farm somewhere on the outer fringes of the British Empire, might naturally be like that, and might nevertheless have the remarkable ability which rumour was beginning to attribute to him.[13]

"Open and friendly in his manner, simple and direct in his judgment of matters," Dale went on. "The friendliness was not returned easily." C. P. Snow, scientist and novelist, who worked with Rutherford years later, wrote, "His voice was three times as loud as any of theirs. His accent was bizarre. It sounded like a mixture of West Country and Cockney."[14] Indeed. A *New York Times* writer, meeting Rutherford a few years later at dinner, asked a friend, "Who was that Australian farmer who sat next to me?" Years after that, meeting the Archbishop of Canterbury, Baron Rutherford of Nelson, which was his title by then, addressed the clergyman as "Your Grice."

That was all to come, though, which shows something of what he must have been like in the self-conscious groves of Cambridge. A classmate, Andrew Balfour, who became an expert on tropical diseases, finally put it this way: "We've got a rabbit here from the Antipodes and he's burrowing mighty deep."[15] The one true foreigner at Cavendish was Paul Langevin, who went on to become one of France's most important scientists. Langevin's room was next to Rutherford's, and when he was asked later whether they were very friendly, Langevin answered, "One can hardly speak of being friendly with a force of nature."[16] 3

It was quite a crew that Thomson put together at Cavendish as the nineteenth century ended. Thomson and then his young men demolished a recurrent scientific myth— one that had surfaced again in the 1870s: that there was nothing left to be discovered, nothing new under the sun. Part of the immutable wisdom of that day, endorsed and believed long before by the greatest of scientists, Isaac Newton, was a kind of billiard ball theory of the atom, which went back to the ancient Greeks. The word itself is from the Greek *atomos*, meaning "indivisible." As Rutherford wrote once, "I was brought up to look at the atom as a nice hard fellow, red or grey in colour, according to taste."[17]

That idea, or theory, was codified in 1804 by an English scientist, John Dalton, who wrote, "1. Small particles called atoms exist and compose all matter; 2. They are indivisible and indestructible; 3. Atoms of the same chemical element have the same chemical properties and do not transmute or change into different elements."[18] So it seemed settled; there was no point in doing research on the atom or a subatomic world, because there was no such thing.

All that began to change, in ways still mysterious, in the year that Rutherford came to Cavendish. On December 28, 1895, Wilhelm Roentgen of the University of Würzburg in

Germany published "On a New Kind of Rays," a paper describing experiments that he had done in the previous two months. Science is a race, and across Europe and in the United States, dozens of physicists were on the same track, observing the effects of electricity passing through different gases at different pressures. Strange and shifting lights were produced inside what they called "cathode tubes," which were gas-filled vessels with positive (anode) and negative (cathode) terminals at either end.

Something was happening as cathode rays shot from one end to the other, but no one yet knew what that something was or what it meant. Perhaps it meant nothing or something useless. Roentgen, who often used the unfamiliar equipment of peers and colleagues, wrapped dark cardboard around part of a tube to prevent light from escaping one day in November 1895. He turned off the lights in his laboratory to check the light seal, and he happened to notice that, as he charged the tube, shimmering light was coming from a fluorescent (barium platinocyanide) screen that he planned to use later. The screen was on a bench more than three feet away from his apparatus. An unknown ray must be coming through glass, an "X-ray" he called it because he did not know what it was. Testing its properties over almost sleepless days and nights, Roentgen discovered that, whatever it was, it could pass through flesh and produce a picture of the inside of his wife's hand, showing the bones and a ring.

Then, two years later, in three "cathode ray" experiments, J. J. Thomson himself discovered that the rays were made up of tiny particles that he called "corpuscles." We now call them "electrons," the negatively charged particles that orbit the positively charged nuclei of atoms. Thomson, however, did not understand that and concluded that atoms were simply clusters of his corpuscles.

Like men of science everywhere, Rutherford was stunned by what he read—and saw. He was still a spectator at the races, but he realized that rays that penetrated matter would inevitably lead to new discoveries and theories of what matter was. "Every professor in Europe is now on the warpath,"[19] he wrote home to Mary Newton in Christchurch. He seemed bored with his radio work. In rather spectacular demonstrations, he had transmitted signals more than half a mile through Cambridge's stone buildings, and there was great interest in finding a way to use such signals in ship-to-shore communications when heavy fog hid the glow of lighthouses. Some scientists, astonished at those demonstrations, believed that Rutherford's work in New Zealand and then at Cavendish was actually ahead of the work being done by Guglielmo Marconi, the Italian scientist-inventor who eventually won the wireless race.

For whatever reason—perhaps he was tired of repeating the sound wave experiments, which may have seemed to him a question of nothing more than sending the same signals farther and farther through thicker and thicker walls— Rutherford was obviously influenced by the fact that radioactivity was the hotter field, attracting most of the world's greatest scientists, beginning with Thomson. But there was also a certain kind of very British pressure being put on him: real scientists were not supposed to make money! Among other things, the great British scientists did not file for patents to protect their work. Mere engineers and inventors did that. If Rutherford followed Marconi, the argument went, he might become rich, but he would be nothing more than an engineer. Fortunately for science and for Marconi, Rutherford did not pursue the engineering problem of improving his apparatus and adapting it to practical use. With Thomson's encouragement, he turned to the more fundamental problem of the passage of electricity through gases. 3

Ironic. Then and for years later, Rutherford's letters to Mary and his mother were punctuated by the anguish of his genteel poverty, quoting the prices of everything he had to buy and the trials of postponing marriage because he could not afford to support a wife. Of necessity, most of his recreation was free. He strained the university libraries to bring him volume after volume of history and biography, becoming something of an expert on the campaigns of Alexander the Great and Napoleon. He walked the countryside, shocked at the dilapidated thatched-roof cottages that dotted the landscape as they had for hundreds of years. "You can't imagine how slow-moving, slow-thinking the English villager is," he wrote to his mother. "He is very different to anything one gets hold of in the colonies."[20]

Rutherford was also something of a prude, writing to Mary after Christmas dinner in 1896, "Some of the dresses were very décolleté. I must say I don't admire it at all. Mrs. X, wife of a professor, wore a 'Creation' I daresay she would call it, which I thought very ugly, bare arms right up to the shoulders, and the rest to match. I wouldn't like any wife of mine to appear so, and I'm sure you wouldn't either."[21]

Whatever he was thinking about his own future, Rutherford quickly put his radio work on hold—though he would occasionally return to it—when J. J. Thomson asked him to join in his radioactivity work. The director saw not only talent and energy in Rutherford, but ambition. The younger man

was burning to succeed, as he told Mary in a letter home in
October 1896: "I have some very big ideas which I hope to try
and these, if successful, would be the making of me. Don't be
surprised if you see a cable some morning that yours truly has
discovered a half a dozen new elements . . . The possibility is
considerable, but the probability rather remote."[22]

The director, Thomson, had been at it for twenty years and
now thought that Roentgen's X-rays might be the tool he
needed to study the production of invisible "ions" (from the
Greek word for "wanderers") and that this young fellow from
New Zealand might be the man to help him do it. Rutherford
jumped at the chance. He loved the work of devising experiments to follow ions, which are atoms or groups of atoms that
carry electric charges because they have gained or lost
electrons. 3

Rutherford loved the work, a first peek into what might be
a subatomic world, made possible because electricity could be
traced. Experiments could follow the charged particles
describing their paths and energy in new formulas. "Jolly little
buggers," he said. "One can almost see them."[23] Only a few
months away from New Zealand, Rutherford was in the new
contest to somehow deconstruct and define matter, the scientific frontier of the day. Scientists everywhere were finding or
looking for experiments and mathematics to describe and
define a world beyond the human eye and microscopes.

Rutherford was now beside Thomson, close to the man
who might turn out to be the winner of the race to define
such phenomena as cathode rays. Thomson assumed—
assumption is at the core of most successful experiments—
that the cathode rays were both electricity and matter. In his
1897 experiments, he proved it. He pushed the cathode rays

around by applying both electric and magnetic forces, finally announcing at a meeting of the Royal Institute in London on April 29, 1897, that the rays were parades of his "corpuscles."

Years of controversy followed, as is almost always the case as scientists build on each others' theories and results, but Thomson was right about the electron, the basis of modern electronics beginning with vacuum tubes. Light could be converted into electric current, or current could be converted into light (the conversion that made television possible).

More than that, Thomson calculated roughly that the mass of his "corpuscles" was a tiny fraction (one eighteen-hundredth) of a hydrogen atom, then the smallest thing known to man. Although he did not use exact words, Thomson was suggesting that atoms were not indivisible. His corpuscles, which he could not yet define, might be particles chipped off of atoms by the energy of electricity and of collisions with other particles. Could this be true? Were there subatomic particles? Some thought the Cavendish director was joking. Thomson himself said, "The assumption of a state of matter more finally subdivided than the atom is a somewhat startling one."[24] He did believe it, though, and proposed the first model of an atom made up of electrons. But he got it wrong.

The question that Thomson and other believers in a subatomic world had to answer then was this: What does an atom look like? Working with Lord William Kelvin, the grand old man of British science at the time, Thomson refined his model of atoms made up solely of electrons and came up with what was called the "plum pudding" model. The atom was a sphere of positively charged something (the pudding) with the negatively charged electrons (bits of plum) scattered about inside. During the first ten years or so of the twentieth century the Thomson–Kelvin model became accepted. But as that decade was ending, new experiments seemed to reveal that the atom was mostly empty space. And the most important of those experiments were done by the man whom Thomson always considered his best student: Ernest Rutherford.

CHAPTER 2

The Rutherford Physics Building on the campus of McGill University in Montreal, Canada, was built in 1977. On a winter day, the lobby of the building is filled with a pyramid of heavy coats, hats, and scarves; and as classes end, students appear from every direction, grabbing their clothes and putting them on before braving the outdoors in that northern place. As they leave, the mound disappears, revealing a bust of the man for whom the building is named, Ernest Rutherford, who came there from the Cavendish Laboratory at Cambridge in 1899, teaching and doing research in Montreal for nine years, until 1908.

The world of physics and physicists was tiny at the beginning of the twentieth century when Rutherford was offered a professorship by the trustees of McGill. Actually he was chosen by J. J. Thomson. In Europe or North America, the Antipodes or Japan, when scholars talked of physics in those days—or when they were looking for new professors—they were usually talking about going to Thomson and Cavendish for recommendations. Many scientists, especially chemists, thought of physics as a rather exotic new discipline, a mix of chemistry, mathematics, and plain old tinkering. As far back as the thinkers and tinkerers of ancient Greece, what we call physics was usually called "natural philosophy"—attempts to figure out what nature was, what the world was made of, and why it worked the way it does. Well into the twentieth century, the *Encyclopaedia Britannica* routinely carried up to fifty pages on chemistry and none on physics.

So it was natural in 1899, when McGill decided to fill the position of chair of experimental physics, the Macdonald Professorship, that the chairman of the two-professor department went to Cambridge to ask Thomson whom he would choose. "Rutherford," said Thomson. His assistant was then twenty-six years old. "I have never had a student with more enthusiasm or ability for original research than Mr. Rutherford," began Thomson's formal recommendation. "I am sure that if elected he would establish a distinguished School of Physics at Montreal."[1]

Rutherford himself was not so sure. "I am only a kid for such a position," he wrote home.[2] He thought he was already working in the best department on the planet. North America, including the United States, was still considered something of an outpost for true research—as opposed to the invention and commercial exploitation done by men like Thomas Edison and Alexander Graham Bell. But there were other considerations, beginning with the twenty-five-hundred-dollar annual salary—enough to get married. It was also enough to pay back the money he borrowed to get to Cambridge, the money he liked to describe, in the language of the frontier, as his "grubstake."

Cambridge men had made fun of words like that. And Rutherford was bothered about the way he was seen and treated there—at a university that famously saw all outsiders as lower forms of life. Among other slights, the young Rutherford was never named a fellow of his college, Trinity College, a prestige he coveted because it provided a campus apartment and a lifetime stipend of 350 pounds sterling per year. He hid his hurt and anger from everyone but his fiancée back home in Christchurch, writing,

> As far as I can see my chances for a Fellowship are very slight. All the dons practically and naturally dislike very much the idea of one of us getting a Fellowship, and no matter how good a man is, he will be chucked out . . . I think it would be much better for me to leave Cambridge, on account of the prejudice of the place, I know perfectly well that if I had gone through the regular Cambridge course, and done a third of the work I have done, I would have got a Fellowship bangoff . . . One has to face the situation squarely and not look always on the rosy side.[3]

There was also the attraction of McGill's new Macdonald Engineering Building, donated by William Macdonald, a Montreal tobacco millionaire who hated smoking, viewing it as a disgusting habit.* Macdonald wanted Canada recognized

---

* There was a double irony in the relationship between Macdonald and Rutherford, who was a heavy smoker. Students and anyone else who got near Rutherford considered his pipe a dangerous weapon, a portable volcano. After coming to McGill, whenever Rutherford heard that Macdonald was on campus, he hustled through the laboratories opening windows to blow out tobacco fumes. In addition, Mary Newton was from a family crusading

as a significant part of the research world, and the building he financed was immediately considered one of the most advanced science buildings in the world. It was new, modern, especially compared to the facilities in the sooty Cavendish building on Free School Lane, an alley really, in Cambridge. Besides, Rutherford would be in charge in Montreal, and the position there emphasized research over teaching. Professor Rutherford's teaching was always more inspirational than comprehensive. Only the best students could follow him into the tangents he was likely to take as his own mind began wandering into uncharted territory.

After ten days of consideration, Rutherford decided to make the move to Canada, writing this to Mary about his new income: "Rejoice with me, dear girl, for matrimony is looming in the distance . . . I am expected to do a lot of original work and to form a research school in order to knock the shine out of the Yankees."[4]

That he did. Room 101 of the Rutherford Building is a small private museum put together in honor of the Physics Department's most celebrated member.[5] Along with photographs of the great man, his papers, and awards from gov-

---

for prohibition of both alcohol and tobacco, and Rutherford had promised his faraway fiancée that he would not smoke—at least, that is, until August 1896, when he wrote this to Mary: "A good long time ago I gave you a promise I would not smoke, and I have kept it like a Briton but I am now considering whether I ought not, for my own sake, to take to tobacco in a mild degree . . . When I come home from researching I can't keep quiet for a minute and generally get in a rather nervous state from pure fidgeting . . . Every scientific man ought to smoke as he has to have the patience of a dozen Jobs in research work." (Arthur Stewart Eve, *Rutherford* [Cambridge: Macmillan, 1939], 39.)

ernments and institutes around the world, there is an extraordinary collection of the equipment Rutherford used in
experiments during his years in Canada. Visitors are
impressed by how compact, how small, the stuff is; much
could be called handheld. From a slight distance some of the
apparatus looks like music boxes, telescope parts, flasks, and
small brass boxes—including the apparatus he used in
radioactive "decay" experiments.

The idea of those experiments, conducted with the vital
assistance of Frederick E. Soddy, a demonstrator in the
Chemistry Department, was to measure the speed and mass
of the "emanations" from radioactive materials. What he discovered then, we now call "half-life." Rutherford and Soddy
recorded that the radioactivity of an element—thorium in
this case—declined in a geometrical progression with time. 3
Thorium radioactivity was reduced by half in sixty seconds,
by half again in the next sixty seconds, and so on. That new
concept made it possible to accurately calculate the age of a
rock in the desert—or of the Earth itself.

It is hard to imagine that so much came from so little,
particularly in physics, a field that now depends on such
tools as particle accelerators that shoot subatomic matter
through miles of tubes and channels, speeding up those particles to bombard and penetrate other matter to discover its
true nature. I was struck more than once by a story told me 3
at McGill, describing Rutherford blowing through a small
tube, a glass straw, to push away gases from the radioactive
thorium in an attempt to discover whether spontaneous
emanations came from a sample's surface or from its subatomic structure. Room 101 dramatizes the idea that nothing in physics was the same after Rutherford—and that he

was the last great "tabletop" physicist. At least in the study of particle physics, he was the last of the great hands-on experimenters. After Rutherford's work as a young man, science, sometimes under his own leadership, moved on to giant testing and measuring machinery and from experimentation to mathematical theory. There was Rutherford and there was Albert Einstein—considered equally great men of science in the first half of the twentieth century—and then there was Niels Bohr, a student and friend of Rutherford who essentially moved nuclear physics from tabletop to paper and pencil or blackboards. 3

At McGill, Rutherford's work produced sixty-nine scientific papers, sometimes coauthored with junior colleagues and graduate students.3His first major paper there, published in early 1900, concerned the source and amount of the energy emanating from uranium and other radioactive materials; it concluded, "It is difficult to suppose that such a quantity of energy can be derived from regrouping of the atoms or molecular recombinations on the ordinary chemical theory."[6] 3

Those words and works were judged eighty years later by the most comprehensive and thorough of Rutherford's many biographers, a British science writer, David Wilson: "The first glimpse, the first implication of what we all know now as 'atomic energy.'"[7] In other words, Rutherford knew already that the huge scale of the energy he was detecting in the emanations was the energy that held atoms together.3 That glimpse, Wilson continued, also illustrated Rutherford's methodology and his genius: "A problem, often an anomalous result, is seized upon; it is worried by that rather slow, but very powerful mind until an answer emerges; and then in a

furious burst of laboratory work the man who all acknowl-
edge to be the greatest experimental scientist of the century
eliminates every other possible explanation and rams his
'intuition' down the throats of every other scientist by
irrefutable evidence."

The *Times* of London would one day describe those quali-
ties in this way: "Rutherford was great rather than clever. His
intellectual machinery was not dazzling: he made no first
impression of subtlety but rather of intense enthusiasm. His
well-built but seemingly not exceptional intellect appeared as
if brilliantly illuminated by a tremendous inner light. His colleagues were startled by the brilliant illumination of the ideas
in his mind, by the pervading clarity and light."[8]

The *Manchester Guardian* described him differently in an
evaluation of his writings: "Beautiful experiments revealing
matters of principle follow one another wondrously, and the
reader says to himself, 'Well, that was brilliantly done; it must
be his finest.' Then on the next half-page there is another of
an entirely different kind, equally good. The procession con-
tinues, page after page, experiments any of which would have
been sufficient to cause the maker to be noticed and approved
in the history of physics. The simplicity of conception, the
cleanness and rapidity of execution were characteristics of
Rutherford's experiments. He seemed almost unable to make
mistakes."[9]

Rutherford, a very young man then in a far place, did
indeed see himself both in collaboration and in competition
with "every other scientist," saying in a letter to his mother
early in 1902, "I have to keep going because there are always
people on my track. I have to publish my present work as
rapidly as possible to keep in the race. The best sprinters on

this road of investigation are Becquerel and the Curies in Paris."[10]

Nineteen of Rutherford's McGill papers were coauthored by Soddy. The young men—Rutherford was then twenty-nine and Soddy was twenty-three—got to know each other in a debate in March 1901, "The Existences of Bodies Smaller Than an Atom"; the chemist's case was entitled "Chemical Evidence of the Indivisibility of the Atom." Soon enough, Soddy realized he was wrong, and that October Rutherford invited the young chemist to join him in his studies of the properties of the thorium emanations. "My recollection of him," Soddy would write a few years later, after the two men had had some differences, "was of an indefatigable investigator, guided by an unerring instinct for the relevant and the important, and as an unequalled experimentalist seeing, amid all the difficulties, the simplest lines of attack . . . I came fully under the influence of his magnetic, energetic and forceful personality, which at a later date was to cast its spell over the whole scientific world."[11]

The man from New Zealand could light fires in other men. The more excited he became, the more excited they became. McGill's resident antiscientist, a classics professor named James McNaughton, who had been known to use the word "plumbers" to describe his scientific colleagues, watched Rutherford in action one day and then wrote,

> We paid our visit to the Physical Society . . . We found that we had stumbled upon one of Dr. Rutherford's brilliant demonstrations of radium. It was indeed an eye-opener. The lecturer himself seemed like a large piece of the expensive and marvellous substance he was describing. Radioac-

tive is the one sufficient term to characterise the total impression made upon us by his personality. Emanations of light and energy, swift and penetrating, cathode rays strong enough to pierce a brick wall, or the head of a Professor of Literature, appeared to sparkle and coruscate from him all over in sheaves. Here was the rarest and most refreshing spectacle—the pure ardour of the chase, a man quite possessed by a noble work and altogether happy in it.[12]

Rutherford was indeed a charismatic and compelling showman; some of his lines were so good that they sounded as if they had been rehearsed. That personality began attracting better faculty and students than McGill had ever hosted, beginning with Soddy and, in 1905, Otto Hahn, a German chemist who spent his postdoctoral year with Rutherford. But evaluations of Rutherford's actual teaching were quite mixed; he had more than a few problems with ordinary students. At one point his department chairman at McGill, John Cox, pulled him aside to say, "Calculus is fine in its place, a useful thing to know. But don't you think, that it is going too far to expect students who have not yet studied calculus to be able to apply it to physics?"[13]

The laboratory, though, was an entirely different world—Rutherford's world. There he could understand, even "see," things that other men could not. Together Rutherford and Soddy attacked the thorium emissions. The mineral was emitting something (or "emanating" to use Rutherford's word), and the amount and speed of the emissions were not affected by temperature, which they would have been if they were the product of ordinary chemical reactions. The emission, they discovered by process of elimination, the essential part of

Rutherford's methodology, was an inert gas that they called "thoron," which was later discovered to be an isotope* of the inert radioactive gas radon, which the Curies had identified as an emission of decaying radium.

With papers suggesting such discoveries flying from the stubby pencils jamming Rutherford's suit pockets, some McGill professors worried that the young man in their midst was going too far, too fast—especially when the Curies openly disagreed with the idea that "emanations" came from the sub-atomic structure of radioactive matter. Pierre Curie was certain that those emanations, or emissions, were produced from the environment outside radioactive material. If Rutherford was wrong in his conclusions, some colleagues argued, the Canadian university would become a laughingstock. This time, Cox was Rutherford's defender, predicting that one day Rutherford would be recognized as the greatest experimental scientist since Michael Faraday, the definer of electricity and electromagnetism. As for Pierre Curie, he repeated Rutherford's experiments and then issued this statement:[†] "I have come to agree with Mr. Rutherford's manner of seeing."[14]

---

* An isotope is a form of an element with identical chemical properties but different physical properties—mass, for example. Specifically, an isotope has the same atomic number but a different atomic weight.

[†] Years later a McGill professor, A. Norman Shaw, wrote of the dispute in the December 1937 issue of the *McGill News*: "When Rutherford was working on the detection and isolation of numerous members of the radium family and developing the theory of the disintegration of matter, there were several occasions when colleagues in other departments gravely expressed the fear that the radical ideas about the spontaneous transmutation of matter might bring discredit on McGill University! At one long-remembered open meeting of the McGill Physical Society he was criticised in this way

Whatever the name of the gas was that Rutherford and Soddy pursued—"thoron" or "radon"—it was a new element. *3* Rutherford and Soddy had created one element from another. "Rutherford, this is transmutation: the Thorium," said Soddy, "is disintegrating and transmuting itself into an argon gas."[15]

"For Mike's sake, Soddy, don't call it transmutation," said Rutherford with a huge laugh, presumably thinking of his doubting colleagues. "They'll have our heads off as alchemists . . . Make it transformation." With that, Rutherford began to march around the lab, doing what he usually did at moments of happy, engaged triumph: he sang "Onward, Christian Soldiers" at the top of his voice—a voice foreign to tune or key but loud enough to move beakers, if not mountains. *4*

The young men, who worked together for just eighteen months, wrote of their results and conclusions in 1902 under the grand title "The Cause and Nature of Radioactivity," saying, "Radio-activity is shown to be accompanied by electrical charges in which new types of matter are continually being produced."[16] So it was "alchemy," not the ancient and fantastic dream of turning lead to gold, but alchemy all the same. It had never knowingly been done before.

It is all still the stuff of legend and anecdote at McGill. Rutherford, they say, was walking across the campus when he ran into Frank Dawson Adams, a professor of geology, and asked, "How old is the Earth supposed to be?" "One hundred

---

and advised to delay publication and proceed more cautiously. This was said seriously to the man who has probably allowed fewer errors to creep into his writings and found it less necessary to modify what was once announced than any other contemporary writer." ("Rutherford at McGill," *McGill News*, December 1937, p. 15.)

million years," said Adams. Reaching into his pocket, Rutherford pulled out a hunk of pitchblende.* "I know for a fact that this piece of pitchblende is 700 million years old," Rutherford said to Adams that day and walked on, chuckling.[17]

So the legend of Rutherford lives on in Room 101, the private McGill museum. There are the vacuum chambers, brass plates, beakers, makeshift batteries, glass tubing, wire, string, and the red Bank of England sealing wax that Rutherford actually used, soldered, wound, and connected in exhibits, almost exactly as they were the last day he used each of them. Usually such devices are long gone, because the equipment of Rutherford's day, generally handmade with the help of creative and resourceful laboratory foremen, was almost always "cannibalized"—dismantled and stored to be reconfigured and used for new experiments by other men and women.

These devices are still intact today because one of Rutherford's former students, another physics professor, Howard Barnes, realized that his mentor was an original, likely to become the most famous professor the school ever had. When Rutherford accepted a new position back in England in 1907,

---

* Pitchblende is a brown and black ore that was extracted for centuries from mountains in the north of Austria, a region that became Czechoslovakia after World War I. A waste product for most of that time, pitchblende took on some value late in the eighteenth century, when it was discovered that it contained the element uranium, which was used to fix the color of ceramic materials. Then, at the end of the nineteenth century, Marie and Pierre Curie discovered that pitchblende also contains elements that came to be called radium. The uranium and radium were, of course, "radioactive"—emitting some kind of energy waves or particles, but no one knew that. No word to describe that phenomenon existed until July 13, 1898, when the Curies coined the term "radioactivity" in describing their theories.

Barnes packed up the experiments and locked them in a closet that was not opened for another thirty years. The museum includes the so-called hundred-thousand-dollar desk that Rutherford used at home—an ordinary wooden desk that cost a small fortune to decontaminate.

Like other researchers of those days, the professor sometimes carried radioactive samples in his pocket and tossed them into a drawer when he got home. The first thing Rutherford had done when he decided to take the position at McGill was order supplies of uranium and thorium, expensive stuff, sent from Germany to Montreal so that he would not lose research time as he moved three thousand miles away from the intellectual centers of Europe. In 1902, Marie Curie, who knew Rutherford only through reputation and their extensive, sometimes contentious scientific correspondence, generously sent him a small sample of radium in solution—as she did with other researchers—making possible new experiments leading to new discoveries about radioactivity.* The McGill museum was finally opened in 1967 and was moved to the Rutherford Building in 1993.

3

---

* Both Marie Curie and, even more so, her husband, Pierre, suffered badly from radiation poisoning. In a weakened state, Pierre was run over and killed in 1906 by a horse and wagon. Marie died of the poisoning in 1934. Rutherford was not as casual as the Curies in handling radioactive material, perhaps because, luckily, he never had radium in the quantity and power that the Curies handled. Once, while demonstrating an experiment at Dartmouth College, Rutherford folded a piece of paper to use as a funnel to pour some radium salt into a tube. Gordon Ferrie Hull, a physics professor there, used that single sheet of paper as Dartmouth's principal radium source for almost forty years. Whatever the reasons, Rutherford never suffered symptoms of radiation poisoning. 4

As for matrimony and Mary Newton, Rutherford did not move as quickly as he often did in the laboratory. After his arrival in Canada in September 1898, he wrote to his Mary, "I have been collecting data since I have been here and I have come to the conclusion that I will have to stay here a year before I go out to fetch you. I start at about zero as regards cash and it would hardly be right for me to fix up before I have settled my bill at home . . . I really think, my dear, we will really have to postpone our partnership for eighteen months from now."[18] 4

Ernest and Mary finally married in the summer of 1900. Their only child, Eileen Mary, was born in Montreal in March 1901. The couple never showed affection in public, and more often than not Mary did not travel with Ernest, but Rutherford was obviously dependent on his Mary, even when she needled him in front of other people—something that happened with regularity. When Soddy was first invited to dinner at the Rutherfords and his host was talking about a paper that Mary had written out for him, she said, "You can't imagine how hard it was to make heads or tails of the original. Ernest never was an orderly person—except in his laboratory. His room at home—he used to live with us in Christchurch—was indescribable. Books and papers everywhere. I know I can't change him but I hope to bring things under control."[19]

To her husband, Mary said, probably more than once, when he wanted her to travel with him or come to the banquets and ceremonies of academia—events he loved—"I am not a society woman. There's no use in trying to make me into one. I love my home and my garden. That's where I belong."[20]

The new bridegroom and father reeled out one astonishing

experiment after another in Canada. The museum in Room 101 quietly celebrates new measuring techniques, the study of emissions from thorium and radium defining the nature and properties of two of the three emissions of radioactivity—he discovered both alpha and beta rays—and the decay experiments leading to the calculation of the age of the Earth. In the end, Rutherford had proved the hypothesis of atomic disintegration as the explanation for radioactivity—an idea accepted without question today. One of his later students, C. P. Snow, the great popular chronicler of the scientific mind and scientific men, wrote of Rutherford, "By thirty, he had already set going the science of nuclear physics—single-handed, as a professor on five-hundred pounds a year, in the isolation of late Victorian Montreal."[21]

The boy from the frontier with an accent strange even for New Zealanders (as a boy the words he heard were spoken with a Scottish burr) seemed to be at a height of both his powers and his energy. In the third paper of their eighteen-month partnership, Rutherford and Soddy said not only things that had never been said before, but some that had never been thought before, as if these two men could see into a new dimension and into the future as well.

In 1902, after the thorium X experiments with Soddy, Rutherford wrote, "I believe that in the radioactive elements we have a process of disintegration or transmutation steadily going on which is the source of the energy dissipated in radioactivity."[22]

In 1903, with Soddy, Rutherford wrote,

Radioactivity, according to present knowledge, must be regarded as the result of a process that lies wholly outside the sphere of known controllable forces . . . All these considerations point to the conclusion that the energy latent in the atom must be enormous compared to that rendered free in ordinary chemical change. Now the radio-elements differ in no way from the other elements in their chemical and physical behaviour. On the one hand they resemble chemically the inactive prototypes in the periodic system very closely, and on the other hand they possess no common chemical characteristics which could be associated with their radioactivity . . . Hence there is no reason to assume that this enormous store of energy is possessed by the radio elements alone.[23]

In other words, they told a startled scientific community that chemistry could never find anything smaller than the atom because the energy bonding together subatomic matter was of a different order than anything previously known or imagined.

Rutherford made two jokes that year—if he was joking— that have resonated through time:[24]

- "Could a proper detonator be found, it's just conceivable that a wave of atomic disintegration might be started through matter which would indeed make this old world vanish in smoke."
- "Some fool in a laboratory might blow up the universe unawares."

Soddy, who eventually gave up science to campaign for world peace, later wrote, "It is probable that all heavy matter possesses—latent and bound up with the structure of the atom—a similar quantity of energy to that possessed by radium. If it could be tapped and controlled what an agent it would be in shaping the world's destiny! The man who put his hand on the lever by which a parsimonious nature regulates so jealously the output of this store of energy would possess a weapon by which he could destroy the earth if he chose."[25]

Not all of this was universally or immediately accepted. Too many ideas and icons were being shattered. This country boy and his chemist friend were fooling around with the theories of Newton, the creator of classic physics, and with the work of science's greatest living figures, beginning with Britain's old lion, the seventy-nine-year-old Lord Kelvin.

In both 1903 and 1904, Rutherford traveled to Europe in the summer, to proudly accept election as a fellow in the Royal Society, Great Britain's scientific honor society, and then to give several lectures, including the annual 1904 Bakerian Lecture of the Royal Society—an unusual honor for a man just thirty-two years old. But in a fundamental way, Rutherford was back as a young student to defend his extraordinary thesis about the way the world worked and to meet many of the men (and one woman) who knew him only from what he had written.

He must have been doing an unseen Maori victory dance. Arthur Eve, who would become his official biographer, said he seemed always to be riding the crest of a wave. Rutherford answered, "Well, I made the wave, didn't I?"[26] In the summer of 1903, Ernest and Mary Rutherford traveled to Europe so that she could meet more of his friends, including one from his Cambridge days, Paul Langevin, in Paris.

The Rutherfords arrived on June 25, 1903, the day that Marie Curie, who was thirty-six and greatly resented by many of the men of the French scientific elite, won her doctorate at the Sorbonne. Rutherford had corresponded with the Curies, but this was their first meeting. It was a night he talked about his entire life, later writing of the small party, "My old friend, Professor Langevin, invited my wife and myself and the Curies and Perrin to dinner. After a very lively evening, we retired about 11 o'clock in the garden, where Professor Curie brought out a tube coated in part with zinc sulphide and containing a large quantity of radium in solution. The luminosity was brilliant in the darkness and it was a splendid finale to an unforgettable day. At the time we could not help observing that the hands of Professor Curie were in a very inflamed and painful state due to exposure to radium rays."[27] Madame Curie did not seem concerned, saying that if radiation could attack healthy skin in that way, it might be useful for attacking the unhealthy cells of cancer and other diseases. Rutherford liked her and her style. As always, she was dressed simply, in an old black dress, and she rarely spoke, knitting and smiling as the men told each other their stories.

On his 1904 European trip, Rutherford attended to the details of the publication by Cambridge University Press of his first book, *Radio-activity*. The first paragraph was as dramatic and heroic as the times in laboratories around the world:

> The close of the old and the beginning of the new century have been marked by a very rapid increase of our knowledge of that most important but little known subject, the connection between electricity and matter. No subject has been more fruitful in surprises to the investigator, both

from the remarkable nature of the phenomena exhibited
and from the laws controlling them ... It has also indicated
that the atom itself is not the smallest unit of matter but is
a complicated structure made up of a number of small
bodies. 3

This was a bold new world. One of the book's most important
reviewers was J. J. Thomson (to whom the book was dedi- 3
cated), who said, "Rutherford has not only extended the
boundaries of knowledge of this subject, but annexed a whole
new province."[28]

Not everyone was so sure. Rutherford was moving so
quickly, leaping hurdle after hurdle of the old physics, that it
was hard to watch him, much less keep up. Lord Kelvin said
he thought he had spent more time reading *Radio-activity*
than anyone else and was still not sure about it. On the night
after he gave the 1904 Bakerian Lecture, Rutherford spoke
again, this time about the age of the planet, a specialty of
Kelvin's:

I came into the room which was half-dark and presently
spotted Lord Kelvin in the audience and realized I was in
trouble at the last part of my speech dealing with the age of
the earth, where my views conflicted with him. To my 3
relief, Kelvin fell asleep, but as I came to the important
point, I saw the old bird sit up, open an eye and cock a
baleful glance at me! Then a sudden inspiration came, and
I said Lord Kelvin had limited the age of the earth *provided
no new source was discovered*. That prophetic utterance
refers to what we are now considering tonight, radium!
Behold! the old boy beamed at me.[29]

Back in North America, the provincial professor found himself something of an international celebrity. A new 558-page edition of *Radio-activity* (the original contained only 382 pages) was in the works, which included his first suggestion that the alpha rays that he continued to study at McGill were actually helium nuclei shot out of radioactive elements, and that the end product of decayed radium and all its descendants was plain old lead. On a more popular front, *Harper's* magazine paid Rutherford 350 dollars for an article entitled "Radium—The Cause of the Earth's Heat." He discussed the power of atomic bonds, and then he had to take on Kelvin again, challenging the old man's estimate that the sun would continue to warm the Earth for five or six million years. More like five hundred or six hundred million years, Rutherford wrote. Newspapers in both England and the United States picked that up under headlines like "DOOMS-DAY POSTPONED!"[30]

The colonial isolation of Montreal did make it harder for many to accept Rutherford's revolution. England, Germany, and France were still the scientific capitals of the time; ideas, results, and formulas shot back and forth across the English Channel in days. North America found out later. Although he had relatively lucrative offers from Yale (which had offered him a chair three times in five years), Columbia, and Stanford and was twice nominated for the directorship of the Smithsonian Institution in Washington, Rutherford wanted to return to Europe. The chance he was waiting for came in a letter in July 1906 from Arthur Schuster, the Langworthy Professor of Physics at the University of Manchester—once home to John Dalton and James Prescott Joule, and considered, after Cavendish, the second most important laboratory in England: *3*

> When we last met I mentioned to you the possibility of my retiring in the near future . . . I have not mentioned the matter to anyone in Manchester yet and I should like you to treat it as quite confidential. But any further steps would be made much easier if I could feel you were ready to step in should a vacancy occur. I really do not know of anyone else to whom I would care to hand over the office.[31]

It was news when Rutherford decided. Under the headline "RUTHERFORD LEAVES McGILL," *The New York Times* wrote on January 13, 1907,

> Montreal, Jan. 12—McGill University is about to suffer a severe loss through the resignation of Prof. Ernest Rutherford, McDonald Professor of Physics since 1898. He has accepted a chair in Victoria University, Manchester, England. *3*
>
> Prof. Rutherford, who is only 35 years old, stands in the front rank among the physicists of the world through his research work on radium, as well as his previous studies and discoveries in connection with wireless telegraphy. *3*

Rutherford came to Manchester on May 24, 1907, after a polite bidding war that involved McGill and Yale over several months. In the process, the British offer was raised from a thousand pounds sterling to sixteen hundred pounds—more than double what he had earned at McGill. He also inherited an enviable laboratory team that included a likable young German technician named Hans Geiger; an unpleasant German named Otto Baumbach, who happened to be one of the world's best glassblowers; and a superb lab "steward" or foreman, William Kay. 3

The bidding was quiet, but Rutherford's entry was not. The new man's public debut was at a science faculty meeting, where he was introduced, walked to the front of the room, smashed his fist on the lectern, and roared, "By thunder!"[1] It seemed that during the months of negotiations between Rutherford and Schuster about the physics job, the head of the Chemistry Department, Howard Baily Dixon, had moved equipment and people into rooms that had once been reserved for physics. Red in the face, Rutherford followed Dixon out of the room, shouting, "You! You are like the fag end of a bad dream!"

Within three weeks, those rooms were being used for Rutherford's latest radioactivity experiments, most of them set up by Geiger. The German described his long working relationship with Rutherford this way: "Rutherford develops the theory, my job and others was to check it."[2] 3

The men of Manchester soon learned what they knew at McGill: Despite his bluff friendliness, Rutherford had a ferocious temper when things went wrong, even if it was his own fault. He was prone to periodic bouts of depression, usually marked by two or three days of scowling and shouting at subordinates foolish enough to confront him when he was down. His saving grace was that he quickly returned to his sunnier self—usually signaled by his off-key choruses of "Onward, Christian Soldiers"—and made the rounds of the laboratory for friendly chats with offended parties. In one case, when he could not find a spectrograph prism, he jumped on the only other person in the room, a 19-year-old undergraduate named Ernest Marsden, and blamed him not only for the missing prism but for almost everything wrong in world science. Learning a few minutes later that another researcher had used the prism and that Marsden had nothing to do with its disappearance, Rutherford returned to the lab, apologized, and spent the better part of an hour asking Marsden about his life and ambitions. 3

As at McGill, Rutherford's new men soon loved him. Some also feared him. The bottom line was inspiration and unwavering loyalty to his own: eleven of his students and associates—including Frederick Soddy, Niels Bohr, and Otto

Hahn—won Nobel Prizes. C. P. Snow and Paul Langevin were only two of the gifted men who closely observed the force of nature that blew in from the Antipodes. One of Rutherford's friends in Manchester was a chemistry instructor named Chaim Weizmann, who would go on to become the first president of Israel and may have been part of the reason that Rutherford would later put enormous time and energy into providing jobs and homes for Jewish scientists fleeing Nazi Germany in the 1930s. Weizmann described his friend this way:

> Youthful, energetic, boisterous, he suggested anything but the scientist. He talked readily and vigorously on any subject under the sun, often without knowing anything about it. Going down to the refectory for lunch, I would hear the loud, friendly voice rolling up the corridor. He was quite devoid of any political knowledge or feelings, being entirely taken up with his epoch making scientific work . . . Rutherford was modest, simple and enormously good natured.[3]

Weizmann, who knew both Rutherford and Albert Einstein well, enjoyed comparing the two giants of early-twentieth-century science:

> I have the distinct impression that Rutherford was not terribly impressed by Einstein's work. Einstein on the other hand always spoke to me of Rutherford in the highest terms, calling him a second Newton. As scientists the two men were contrasting types—Einstein all calculation, Rutherford all experiment . . . There was no doubt that as

an experimenter Rutherford was a genius, one of the greatest. He worked by intuition and everything he touched turned to gold. He had a sixth sense.[4]

They were very different men. Or boys. Someone said they were both like curious children—Einstein the merry boy, Rutherford the boisterous one. They were looking and working in different directions—Einstein looking outward, rather dreamily trying to discover where we came from, and Rutherford drilling deep to discover what we were.*

At a dinner in 1910, Wilhelm Wien, a German who would win the Nobel Prize in Physics a year later, teased Rutherford, "But no Anglo Saxon can understand relativity."[5] Rutherford laughed his big laugh and said, "No! They have too much sense." And often Rutherford said to his researchers, "Don't let me catch anyone talking about the Universe in my department."[6]

Great fun. But Rutherford, whose personal relationship with Einstein was quite good, may have understood more about that great man and his work than he pretended. At a

---

* The scientific arguments of the time between experimentalists and theorists in Europe often came down to comparisons between Rutherford and Einstein. In *Rutherford, Simple Genius* by David Wilson, published in 1983, the author felt compelled to write this footnote on page 83: "One of the difficulties of writing this book on Rutherford is to avoid the appearance of 'knocking' Einstein. Because his great Theory of Relativity embraces so many phenomena that seem incredible to the layman, such as the equivalence of mass and energy, formulated as the famous $E = mc^2$, and because so many modern scientific statements start from the firm basis of Einstein's work, it is not realised how many of the conclusions we associate with Einstein were in fact discovered by other scientists before him."

dinner of the Royal Academy of the Arts in 1932, Rutherford's role was to respond to the toast, "Science!":

> I think that a strong claim can be made that the process of scientific discovery may be regarded as a form of art. This is best seen in the theoretical aspects of Physical Science. The mathematical theorist builds up on certain assumptions and according to well understood logical rules, step by step, a steady edifice, while his imaginative power brings out clearly the hidden relation between its parts. A well constructed theory is undoubtedly an artistic production ... The theory of relativity by Einstein quite apart from any question of its validity, cannot but be regarded as a magnificent work of art.[7]

Rutherford brought more than reputation and prizes to Manchester. He came with the element more valuable than gold: radium. When Rutherford arrived in Manchester, the university had only seven milligrams of the magic element discovered by Marie and Pierre Curie. It was simply too difficult and too costly to isolate significant amounts of radium from pitchblende. The cost to produce a gram—this was 1907—was estimated at more than a hundred thousand dollars, which would make a pound worth 450 million dollars. But because Rutherford was there, Manchester was offered pitchblende from the Curies' stock and was lent 350 milligrams of radium bromide by the Radium Institute of Vienna, an Austrian government committee that controlled the distribution of pitchblende mined in its mountains.

"In New Zealand we don't have money, so we have to think"[8] was one of Rutherford's repeated maxims. He put it to

good use in figuring out ways to take advantage of Baumbach's glassblowing talents. One of those ways was to reuse
the radium in what he called "alpha ray tubes"—vessels less
than one-twentieth an inch in diameter, with glass thin
enough to leak a supply of alpha particles from emitted radon
gas. So instead of using an actual radium source, Rutherford
and other Manchester researchers could substitute the tiny
glass tubes.

Looking for proof that alpha particles were actually the
nuclei of helium atoms, Rutherford had Baumbach craft a
glass cylinder with the thinnest walls possible—there was a
small mountain of broken glass in the lab as the work
proceeded—and then place that delicate tube inside another
thicker glass cylinder. The inner cylinder, the smaller tube, had
walls just a hundredth of a millimeter thick, which meant that
they were thin enough to allow the passage of alpha particles
but thick enough to hold in the radioactive radon gas. Rutherford then pumped the gas into the inner vessel and pulled a
vacuum in the outer cylinder. After six days, he calculated, the
tube should be filled with nothing but alpha particles—at least
that was the plan. And it worked. In a paper titled "The Nature
of the Alpha Particle," sent to the Manchester Literary and
Philosophical Society on November 3, 1908, Rutherford
described the work and the inner cylinder: "The walls of this
glass tube were sufficiently strong to withstand atmospheric
pressure but thin enough to allow the greater number of alpha
particles to be fired through them. Mr. Baumbach succeeded
in blowing a number of such fine tubes for us."

Finally, then, Rutherford was able to study the isolated particles in the outer tube using a spectroscope. He knew what he

was looking for, and when he passed electricity through the outer cylinder, he found it. He had long ago guessed that alpha particles were actually positively charged atoms of helium, but he had never quite said that publicly, because he
3 could not prove it. This time the spectroscope showed the distinctive yellow and green spectrum lines that identified the presence of helium in the gases of the sun. Four days after the British publication, on November 8, 1907, Rutherford's experiment was reported in a full column on the front page of *The New York Times*, under a deck of headlines:

ATOM OF MATTER CAN BE DETECTED
Prof. Rutherford, Expert on Radioactivity,
Makes Successful Experiments

SUBSTANCES TRANSFORMED
Accomplished by Expulsion of "Alpha Particle"
Which Prof. Rutherford Believes Is an Atom of Helium

The report was based on Rutherford's own report to the *Scientific Weekly* in London. Then *The New York Times* wrote,

Professor Rutherford's conclusion is put with characteristic modesty. "It may be of interest to note," he says, "that the experimental results reported in this article lead to experimental proof, if proof is needed, of the correctness of the atomic hypothesis with regard to the discrete structure of matter."

This is truly a modest claim of victory in the greatest scientific battle of the century.

The newspaper quoted a variety of experts and implications, including this from Soddy, who had moved on to the University of Edinburgh: "I am in entire agreement with Prof. Rutherford's conclusions . . . The experiments are beyond all praise."

There was, however, something that the *New York Times* correspondent in London did not know. As Rutherford was sending his report off to the *Scientific Weekly*, he received a telegram from Stockholm, Sweden. He would be awarded the 1908 Nobel Prize in Chemistry "for his researches concerning the disintegration of the elements and the chemistry of radioactive substances"[9]—the nuclear decay work that he had done back at McGill in 1903.

The prize was out of the ordinary in more ways than one, at least for Rutherford. He had always been rather single-mindedly dedicated to the advancement of physics, sometimes mocking chemistry and chemists. "All science is either physics or stamp collecting," he liked to say.[10]

The Nobel Prize presentation speech on December 10, 1908, by K. B. Hasselberg, president of the Royal Swedish Academy of Sciences, began, "Rutherford chose [radioactive rays] as a subject of very thorough investigation, evolving exceedingly exact methods for measuring their intensity, proving the existence of absolutely distinct types of rays (the so-called α-rays and β-rays), establishing the more important characteristics of the two types, and bringing forward, more especially as regards the α-rays, unimpeachable proof of their material nature."[11]

Indeed he had. Rutherford began his acceptance speech with a small joke, saying that he had seen many transformations in laboratories but the quickest one was the Nobel committee's "instantaneous transmutation" of him from physicist to chemist. The audience of past laureates and other dignitaries laughed, but many of them were whispering about how young Rutherford looked. At thirty-seven, he still looked like a boy, at least to them. He was big but trim; his hair was still blond. He was immediately serious as he gave his own speech, a long one generous to other researchers as he related the history of radioactive research and experimentation. His title was "The Chemical Nature of the Alpha Particles from Radioactive Substances."

Rutherford ended in a big way, announcing what he and Geiger had found using their little double tube of glass:

> Considering the evidence together, we conclude that the α-particle is a projected atom of helium, which has, or in some way during its flight acquires, two unit charges of positive electricity. It is somewhat unexpected that the atom of a monatomic gas like helium should carry a double charge. It must not however be forgotten that the α-particle is released at a high speed as a result of an intense atomic explosion, and plunges through the molecules of matter in its path. Such conditions are exceptionally favourable to the release of loosely attached electrons from the atomic system. If the α-particle can lose two electrons in this way, the double positive charge is explained.[12]

Almost explained. The alpha particle was a helium nucleus stripped of its orbiting electrons. But no one then knew what an atom looked like, and both the idea of the nucleus and the word itself did not exist. Looking into the atom, "seeing" the subatomic world—that's where Rutherford was going next.

First, though, there were banquets and celebrations, beginning in Stockholm and moving on to Germany and the Netherlands—"Lady Rutherford and I had the time of our lives," said the new laureate[13]—ending in February with a dinner at Manchester. The speaker was the first winner of the Nobel Prize for Physics, given in 1906 for the discovery of the electron, Rutherford's great mentor and champion, J. J. Thomson. Oddly enough, he began with an apology:

> Professor Rutherford never received the credit he should have had for his work in connection with radiotelegraphy in 1895. His success was so great that I have since felt some misgivings that I persuaded him to devote himself to the new department of physics that was opened by the discovery of Röntgen rays . . . The discovery at McGill of the emanation of Thorium was a stroke of genius, because the new gas had to be endowed with properties not recognized as belonging to any such known substance, as it is a gas which exists for a few minutes only; half of what there is of it always disappears in less than a minute! Of all the services that can be rendered to science the introduction of new ideas is the greatest.[14]

Rutherford responded that no apologies were necessary: "The progress of science [is] like the progress of man going through a swamp, with islands of firm earth in between. The advances would be very slow from week to week, but at the end of a year it was great, and at the end of ten years it was enormous. In physics there had been a revolution since 1896 . . . there was nothing more interesting than watching the progress."[15]

CHAPTER 4

Rutherford loved to tell stories. He rambled on at tea with his researchers each afternoon in his laboratories. Sunday teas at home usually turned into what he called "my powwows" in his study. He told one story after another on long car rides after he bought a sixteen-horsepower Wolseley-Siddeley with part of his 7,680-pound Nobel award, and he held forth in the world's most prestigious lecture halls as he became more and more famous. More than anything, he wanted to talk about his work, knowing that talk was one way of thinking things out. When one of Rutherford's many official portraits was ceremoniously unveiled, his wife tartly remarked that she was surprised the artist had caught him with his mouth shut. ꝥ

Rutherford's favorite story was about a few days in May 1909 that changed the world of science. He began it this way: "One day Geiger came to me and said, 'Don't you think that young Marsden ought to begin a small research?' Now I had thought so, too, so I said, 'Why not let him see if any alpha particles can be scattered through large angles?' Two or three days later Geiger came back to me in great excitement, saying, 'We have been able to get some of the alpha particles coming backward' . . . It was quite the most incredible event that has ever happened to me. It was almost as incredible as if you fired a 15-inch shell at a piece of tissue paper and it came back and hit you."[1]

Geiger had his own stories about those days. He and "The Prof," as he always called Rutherford, worked in the basement of the physics building: "I see the gloomy cellar in which he had fitted up his delicate apparatus for the study of the alpha rays. Rutherford loved this room. One went down two steps and then heard from the darkness Rutherford's voice reminding one that a hot water pipe crossed the room at head level, and to step over two water pipes. Then finally in the feeble light one saw the great man himself seated at his apparatus."[2]

Neither Rutherford nor Geiger was sure what had happened in the basement on that day in 1909. The incredible event was the climax of a long series of "scattering" experiments that the two men had begun working on as soon as the new professor arrived at Manchester. Trying to establish the speed and properties of radioactive particles, they used radium or radon as a radioactive source and began "firing" alpha particles into a vacuum chamber. The "gun" was a lead box with a tiny slit that allowed a small number of the particles to propel themselves into the chamber. Or, to be more specific, *some* of the particles, since one of their findings was that one gram of radium released more than thirty-four billion alpha particles per second, most of which were blocked by the lead enclosure.

The target was a small plate coated with phosphorescent

zinc sulfide at the back of the vacuum chamber. They could not see the particles, of course; the subatomic alpha particles were traveling at one-seventh the speed of light. But what they could see, with a microscope looking into the chamber, were "scintillations," tiny and brief sparks of light when the particles struck the phosphorescent screen, leaving marks where they hit.

The tiny hits produced a thin line on the screen, seeming to mirror the slit through which they were "fired." Rutherford, however, knew that the particles could be deflected a degree or so in a couple of ways, by strong electric or magnetic fields or by being passed through matter. At McGill in 1906, he had fired alpha particles through mica and had seen that the sharp line on the screen got a bit fuzzy as particles deviated from the straight and narrow. It was a piece of information that he had filed away somewhere in his brain.

Rutherford was a great one for mulling over aberrations, anything out of the ordinary. That's where the secrets were, in the margins. In his decay experiments at McGill, for instance, he discovered alpha and beta particles not by studying radium and uranium themselves, but by concentrating on the gases that they emitted, the "emanations." Radium and uranium were only tools. Now the alpha particles were Rutherford's tools; he was using them as projectiles to take him inside matter. He was trying to "look" inside atoms, to "see" the atom.

Thus, Rutherford and Geiger began using their apparatus to fire alpha particles through metal foils to see whether and how they scattered. What they found first was that the denser the foil elements were—as judged by their atomic weight— the fuzzier were the phosphorescent lines that they saw through their microscope. Obviously the denser the element was, the more likely it was that the alpha particles would have some interaction—physical or electrical—before passing through and reaching the phosphorescent target. Gold (atomic weight 197) produced about twice as much scattering—fuzziness—as silver (atomic weight 107), and almost twenty times as much deviation as aluminum (atomic weight 27).

Rutherford knew he was onto something. But what? Those early results were limited by several factors, including the fact that it was extremely difficult to evacuate vacuum chambers with the hand pumps of the early twentieth century. The odd molecule of air or speck of dust could and did ruin the best of experiments. The field of vision of the fixed microscope used in counting scintillations was quite limited, raising the possibility that other alpha particles were hitting the target beyond the edges of their vision. That was why Rutherford wanted Marsden, under Geiger's supervision, to build an apparatus that might show wider scattering, if there was any. Finally, because the early experiments caught only a small percentage of the millions of alpha projectiles, Nobel Laureate Rutherford went back to school in 1909. He sat in on every class of a freshman-level course in statistical probability to understand whether the results produced by his limited sample of alpha particles were a valid representation of the entire universe of particles, the counted and the uncounted.

Rutherford's story about his two young men coming back with startling results in two or three days was an obvious exaggeration, and he probably exaggerated his own surprise as well. The apparatus took time to build; its innovation was making it possible for the microscope to move around the circumference of the chamber without breaking the vacuum.

Now they could see wide deviations, but only after sitting in the darkened laboratory long enough to adjust their eyes to use the microscope. That was their job. Rutherford himself did not have either the patience or eyes good enough to do much of the observing, saying, "Geiger is a demon at the work and could count at intervals for a whole night without destroying his equanimity. I damned vigorously after two minutes and retired from the conflict."[3]

The same goes for me. With help and guidance from Kurt Becker of the Physics Department of Stevens Institute of Technology, I re-created the final and most famous scattering exercise in the workshop and then a laboratory of the technical university in Hoboken, New Jersey. I'm not sure I lasted even two minutes; certainly my old eyes were not up to the task at hand. After an hour in the dark, I will never be completely sure that the scintillations I "saw" were real or just wish fulfillment. The young eyes and swift minds of two graduate students, Frank Corvino and Damien Marianucci, however, were able to see and record results. Marsden himself was twenty years old when he was the man on the microscope in that Manchester basement.*

---

* I graduated from Stevens in 1960 with a degree in mechanical engineering. The re-creation of the scattering experiment is described in the introduction of this book. Rutherford's young eyes, Geiger and Marsden, went on to distinguished careers of their own. Geiger became a household name, inventing the Geiger counter in 1928 to replace tired eyes checking for and assessing levels of radiation. Sir Ernest Marsden became a professor of physics at the University of New Zealand and was considered the country's second most distinguished scientist—after Rutherford. Later in life, Marsden wrote that modern safety restraints would have made the scattering experiments "unworkably complicated."

After setting up the apparatus, Marsden and Geiger began to place metal foils between the "gun" and the "target." They began with light metals like lithium and worked up to gold. The gold foil, placed at a forty-five-degree angle to the speeding alpha "bullets," was 6/100,000 of a centimeter thick, which translated into a thickness of perhaps four hundred atoms. Most alpha particles passed through without deviation, many deviated a degree or two, and, the young men reported back to Rutherford, some did fly out at ninety degrees. Rutherford told them to put the scintillation screen, the target, in front of the foil, positioned just out of range of the beam of the particles. They did, and they were stunned at what happened inside their little chamber: one in about eight thousand particles bounced back at an angle of more than 90 degrees, some coming almost straight back, at 150 degrees and more. That was when Rutherford raced to the lab and came up with his line about tissue paper.

Geiger and Marsden—the teenager whom Rutherford had blamed for moving a prism a year earlier—published their results under the title "On a Diffuse Reflection of the Alpha Particles" in the June 1909 edition of *Proceedings of the Royal Society*: "In the following experiments . . . conclusive evidence was found of the existence of diffuse reflection of alpha particles. A small fraction of alpha particles falling upon a metal plate have their directions changed to such an extent that they emerge again on the side of incidence."

"At first we could not understand this at all,"[4] Geiger wrote later. Among other astonishing hypothetical realities, on paper it would take billions of volts to turn an alpha particle. The German went back to experiments on small-scale deflection. Marsden left to study meteorology, then moved on to a position at the University of New Zealand. Rutherford tried a few more scattering experiments over the next few months, but he was busy running a university department and accepting awards. Still, he practiced something he preached. With the twenty-hour lab marathons of his youth behind him, he closed the Manchester labs at six o'clock in the evening, ordering a new generation of students and researchers to "go home and just think."

Rutherford himself thought about the scattering results for more than a year. He understood that the alpha particles, pretty tough customers themselves with almost eight thousand times the mass of electrons, had run into something infinitely more massive. By early December 1910, he had obviously formed a picture in his mind of an atomic "nucleus"—though that word did not yet exist—writing the following in a letter to an American friend and colleague, Bertram Boltwood of Yale, who had taken a one-year sabbatical at Manchester: "I think I can devise an atom much superior to J. J.'s for the explanation and stoppage of alpha and beta particles, and at the same time I think it will fit in extraordinarily well with the experimental numbers. It will account for the reflected alpha particles observed by Geiger and generally I think will make a fine working hypothesis. Altogether I am confident that we are going to get more information from scattering about the atom than from any other method of attack."[5]

The Rutherfords hosted a Christmas dinner the Sunday before Christmas in 1911, inviting friends and colleagues. One of the guests was Charles G. Darwin, Rutherford's resident mathematician and grandson of the author of *The Origin of Species*, who would later write, "One of the great experiences of my life . . . after supper the nuclear theory came out . . . He assumed a central charge in the atom—it was indeed a year or two before it was named the nucleus—repelling the alpha particle according to the ordinary laws of electricity . . . I also recollect that even on that first evening Rutherford was also speculating on how small the nucleus might be."[6] The next morning, Rutherford walked into Geiger's lab, stood there for a moment, and then, as Geiger remembered, "He told me that he now knew what the atom looked like and how to explain the large deflections of the alpha particles."[7]

Rutherford's atom was mostly nothing—a few electric particles wheeling through a vacuum. The central charge, which he did not call the "nucleus" until 1913, was a dense, electrically charged particle surrounded by much smaller and lighter bits of matter orbiting around it somewhat in the manner of planets around the sun. Rutherford did not know whether the central charge was positive or negative, but it was powerful enough to deflect back alpha particles that came close; that was the wide-angle scattering observed by Geiger and Marsden. The relative size of the central charge (the nucleus) he estimated at one-trillionth of an inch, occupying one-billionth of the space called the "atom." It was, then, relatively the size of a pinhead at the center of St. Paul's Cathedral, but it accounted for 3,999/4,000 of the mass of the atom.

Rutherford reported his findings and theory on March 7, 1911, writing cautiously under the title "The Scattering of the Alpha and Beta Rays and the Structure of the Atom" even as he demolished the existing views of what atoms were and how they made the world that men saw:

It is well known that the alpha and beta particles are deflected from their rectilinear path by encounters with the atoms of matter ... There seems to be no doubt that these swiftly moving particles actually pass through the atomic system, and the deflexions observed should throw light on the electrical structure of the atom ... Geiger and Marsden found that a small fraction of α particles incident on a thin foil of gold suffers a deflexion of more than 90-degrees* ... It seems certain that these large deviations of the α particle are produced by a single atomic encounter ... A simple calculation shows that the atom must be the seat of an intense electric field in order to produce such a large deflexion at a single encounter. Considering the evidence as a whole, it seems simplest to suppose that the atom contains a central charge distributed through a very small volume ... In comparing the theory outlined in this paper with the experimental results, it has been supposed that the atom consists of a central charge concentrated on a point."[8]

Two years later, in his third book, *Radioactive Substances and Their Radiation*, Rutherford declared that new experiments showed that the "central charge," orbited by negatively charged electrons, was positive, and then he coined the name for it: the "nucleus." He had "seen" the atom pretty much as we do today. C. P. Snow, who wrote that Rutherford had created the study of

---

* The alpha "bullet" did not actually strike the nucleus of a gold atom. The deflection was due to electrostatic repulsion. When both the particle and the nucleus were discovered to be positively charged, it became obvious that, following the rules of attraction and repulsion, the particle was being repelled by the nucleus.

nuclear physics by the time he was thirty years old, added, "By forty, now at Manchester, he had found the structure of the atom—on which all nuclear physics depends."[9]

Reading of Rutherford's work, Arthur Eddington, the Cambridge astronomer whose work helped confirm some of Einstein's theories, said that Rutherford's discovery was the most important development in science since Democritus, the greatest of the Greek natural philosophers, proposed the existence of the atom in the fifth century BC. But not many other people agreed (or even noticed) at first, accepting the Rutherford atom as just another report from the powerful idea factory in Manchester. The most pertinent response came from Hantaro Nagaoka, a Japanese physicist who had previously suggested that the atom looked something like the planet Saturn, with its mass concentrated in a huge globe orbited by concentric rings of electrons. He wrote Rutherford from Tokyo, "Congratulations on the simpleness of the apparatus you employ and the brilliant results you obtain."[10]

Nagaoka's work is mentioned in several Rutherford papers of the period, but he is best remembered in Rutherford legend for telling the story of Japanese Minister of Education Baron Kikuchi's introduction to Rutherford at Manchester. Arthur Schuster did the honors, and the Japanese minister later asked him, "I suppose the Rutherford you introduced me to is a son of the celebrated Professor Rutherford."[11] Eddington, too, is remembered anecdotally for what he said at a Cambridge dinner party, suggesting that perhaps electrons were a mental concept and did not actually exist. "You have insulted the woman I love," spluttered Rutherford across the table. "Not exist? Not exist? Why I can see them as plainly as I can see that spoon in front of me."[12]

And he could. Rutherford's genius was in making other people see, beginning with his students and colleagues, and then showing the world. But, like almost all scientific "answers," Rutherford's experimental breakthrough raised new questions. One modern standard physics text puts it this way: "Successful as this model of the nuclear atom was in explaining scattering phenomena, it raised many new questions. What is the arrangement of electrons about the nucleus? What keeps the negative electron from falling into the positive nucleus by electrical attraction? Of what is the nucleus composed? What keeps it from exploding on account of the repulsion of its positive charges?"[13] The textbook might have added, "And starting an atomic chain reaction that would blow up the world."

Rutherford realized the problems raised by these questions and the failure of his model to answer them all. Additional assumptions were needed to build on the model, which could never be complete. Science is a cumulative discipline, an ongoing attempt to answer the most basic questions about the universe and mankind—what is it? what are we made of?—which was why Rutherford was such a champion of the speedy and open publication of results. He knew, too, that the most immediate of those questions, once the nucleus was discovered to be positively charged, was why the electrons were not gradually drawn into the nucleus. The Rutherford model seemed inherently unstable. Shouldn't the structure implode?

Then, in a match made in science heaven, Rutherford met a twenty-seven-year-old University of Copenhagen physicist named <u>Niels Bohr</u>, who had come to England to do research under J. J. Thomson at Cavendish. Bohr, who thought his quiet and deep talents were being wasted in Cambridge, attended the annual Cavendish dinner in December 1911. These were wild, happy affairs with bawdy toasts about great men, yelling, and cheering, ending with men like Thomson and Rutherford standing on chairs, crossing their arms, holding hands, and singing, "God Save the King." The principal speaker that year was Rutherford, down from Manchester for the night. Bohr was enormously impressed, excited, and inspired. He was not the first nor the last young scientist to decide to change his life to join Rutherford, but he was probably the most important.

The timing was perfect. The two big men, both former athletes, hit it off immediately—some said that Bohr became the son Rutherford never had—and the young Dane packed up and headed for Manchester, where he practically lived with the Rutherfords. "I received a deep impression of the charm and power of his personality by which he had managed to achieve almost the incredible wherever he worked," Bohr wrote home to the woman he would soon marry. "The dinner took place in a most humourous atmosphere and gave the opportunity for several of Rutherford's colleagues to recall some of the many anecdotes which already were then attached to his name."[14]

The admiration, friendship, and collaboration lasted a lifetime. Geiger and the others were surprised by that, because Bohr was no experimenter; he was the kind of theoretical scientist/mathematician, Albert Einstein among them, that Rutherford sometimes scorned. "No, no, Bohr's different," said the leader of the Manchester tribe, called "Papa" behind his back. "He's a football player!" Rutherford also said that the Dane was "quite the most intelligent chap I've ever met."[15]

Bohr was perhaps the only person in the world who could convincingly stabilize the Rutherford model—on paper. In several months of intense scribbling and talking, Bohr showed, with formulas, how Rutherford's atom could be stable. He began with the balance that could be attributed to classical centrifugal forces and then formulated new concepts of stability using the "quantum" theories of the German physicist Max Planck. It was Planck's hypothesis that light and other energy forms were not continuous rays moving through space, but were actually small packages of radiant energy moving in pulses. Planck's work was yet another attack on classical physics—and it had already been used, in 1905, by another theoretical physicist, Einstein, to explain photoelectric effects.* In what came to be known as the Rutherford–Bohr model of the atom, when outside energy touched atoms— light or radioactive heat, for example—excited electrons absorbed that energy and jumped, made a "quantum leap" to a new orbit. That, along with centrifugal force, was what prevented their descent into the nucleus. 3

"Those were happy days at Manchester," Rutherford said

---

* For his work, Planck was awarded the 1918 Nobel Prize in Physics, an honor that always made him uncomfortable because he was a disciple of classical physics. The use of quantum theory won Einstein the same prize in 1921. British science writer Brian Cathcart, author of an engaging history of the splitting of the atom titled *The Fly in the Cathedral* (New York: Farrar, Strauss and Giroux, 2004), described quanta (as opposed to continuous movement) by using a clever kitchen analogy: "In baking, milk is 'continuous' in the sense that any given amount may be measured out and added to a mixture, while eggs tend to be 'discontinuous' . . . In physics, the classical laws considered light, motion, heat and energy to be like milk rather than eggs; they could come in any quantity, no matter how small. When atoms absorbed energy or gave it out, therefore, they must obey this principle and deal in infinitely variable quantities. Not so, said Planck . . . It was not a case of milk but rather of eggs—or in Planck's terms, quanta" (p. 70).

years later to Geiger.[16] "Papa" attracted and assembled a dazzling tribe of great young scientific talent. One of his "boys,"
Edward Neville da Costa Andrade, an Englishman who came
from the University of Heidelberg in Germany, said, "It was a
Utopia, really, with the Professor in closest touch with all his
research men who, with little thought for their future living,
were eagerly engaging themselves in obtaining results that
seemed remote from any possible practical application."[17]
Rutherford had inherited Geiger and Marsden, but then he
attracted Bohr; B. B. Boltwood of the United States; Georg
von Hevesy of Hungary; Kasimir Fajans of Poland; Marcus L.
Oliphant of Australia; and the Englishmen Andrade, James
Chadwick, Charles G. Darwin, John D. Cockcroft, and Henry
Gwyn Jeffreys Moseley, creator of Moseley's law of atomic
numbers and perhaps the most promising of them all.
Rutherford ended that conversation with Geiger by saying,
"We wrought better than we knew."[18]

The Rutherford–Bohr model, a product of inspired experimentation and imaginative theory, was an end and a
beginning—the beginning of the end of Rutherford's own
sealing wax and string tabletop physics. Classical, hands-on
physics was beginning to give way to blackboards. New experiments breaking into the nucleus would require giant
machines to accelerate and manipulate the classic forces and
bodies first defined and mastered by Isaac Newton, Michael
Faraday, J. J. Thomson, and Rutherford himself.

It could have been the end of Rutherford too. But, in many
ways, he was just getting started, moving now in more than
one direction. The great battleship, the force of nature, continued moving ahead as an experimenter who would continue to astonish, splitting the atom he first "saw." But he was

also spending more time as a mentor and inspiration to a new generation of world-changing scientists, and emerging as a respected and powerful statesman of science.

As a great experimentalist and a hands-on guide to younger men, Rutherford may have had an advantage over the other giant of his generation, the great theoretician Einstein, who essentially worked only inside his own head. Talking about the two men, another Nobel laureate for Physics, Pierre-Gilles de Gennes of France, the 1991 winner, told me, "Some of us overestimate theory, but we need efficient scientists able to do things with their hands, like Rutherford. Yes, I'm a theorist and there is a need to know the basics, a classical musician needs to know the basics of playing the cello. But the fields and the men are different. In mathematics: before you are 35, you burn out. Some novelists are like that. Physics and experimentation are more like painting, you can do it into a ripe old age . . . There is something to be said for the shopkeepers, fortunately one can benefit from experience."[19]

Inspired by the example of J. J. Thomson, Rutherford created a shop that produced dozens of scientists of world renown over forty years. That change of direction, or growth, tends to be confirmed by inspection of his public papers. Young Rutherford wrote as many papers in nine years at McGill as his combined total in thirty exalted years at Manchester and Cambridge.

As he began to grow older—"Ern, you're getting stout,"[20] his wife would tell him—Rutherford continued to work with his hands; there were still masterpieces within his reach, but he also loved the roles of mentor and statesman, the shopkeeper of nuclear physics. He enjoyed it all, especially the pomp of recognition. On New Year's Day 1914, he came home from the lab and said, "Good evening, Lady Rutherford." Their daughter, Eileen, thirteen that year, asked what he meant. He said he was on the King's New Year's Honors List. "Henceforth, young lady, you may address me as 'Sir Ernest.'"[21]

Rutherford liked to tell that story. He also repeated endlessly one about finding himself alone in a waiting room with a man he did not know: "Hello. I'm Lord Rutherford," he said. The other fellow answered, "Hello. I'm the Archbishop of York."[22] Then Rutherford laughed his huge laugh and added, "I don't suppose either of us believed the other."

CHAPTER 5

The writer of Rutherford's most important biography, David Wilson, wrote this in 1984: "The years 1911 and 1912, which saw Rutherford creating the hypothesis of the nuclear atom, also saw him recruiting Niels Bohr and Harry Moseley; if the director of a modern research laboratory had recruited two such men in a life-time of work he would be regarded as a genius for that alone."[1]

Henry Gwyn Jeffreys Moseley, only twenty years old at the time, wrote a letter to Rutherford saying that he doubted his Oxford education would be of much value in Manchester but he dreamed of working in a laboratory. "I would like to be guided entirely by you . . ." he wrote, adding, "I will spend August in Oxford, and then will read up on Radioactivity . . . My present knowledge extends little way beyond your books."[2] Moseley turned out to be a young man of dazzling talent, probably the hardest-working man anyone at Manchester ever saw, defying Rutherford's rules about going home for dinner at six. He was also an old-fashioned British snob and racist, writing home that Rutherford was "the son of a New Zealand flax farmer who possessed neither languages or culture."[3] As for the students he taught, Moseley said, "I was disgusted to find a large proportion of coloured students .⸫. These seemed to include Hindoos, Burmese, Jap, Egyptian and other vile forms of Indian. Their scented dirtiness is not pleasant at close quarters."[4]

However, Moseley had obvious brilliance, and he was the only member of Rutherford's "tribe" to force the boss to publicly admit to a mistake in his work. Early in 1912, Rutherford made a series of mathematical errors in a paper on the origins of beta rays, and in the December 1912 issue of *Philosophical Magazine* he wrote a correcting letter that included this statement: "Mr. Moseley drew my attention to the fact, which I had overlooked, that according to the Lorentz-Einstein theory ⸫. ."

More than that, Moseley, at the age of twenty-six, published two papers in just six months in 1913 and 1914, following up on Rutherford "guesses" or "hunches." In the same period, Moseley established the modern law of atomic numbers, which included a sentence famous in nuclear physics: "There is a fundamental quality in the atom which increases by regular steps as we pass from one element to the next. This quality can only be the charge on the central positive nucleus."[5] In a series of experiments establishing that X-rays have characteristic spectrum lines and that those lines differ for each element in an orderly manner, Moseley established that, ignoring atomic weights, the elements could be ordered by the positive charge of the nuclei, from hydrogen with a defining positive charge of one, through uranium with ninety-two positive charges in its nucleus. In other words, the properties of the elements that made up the world were determined only by the positive charge of each nucleus. Among the beliefs superseded by that work was Rutherford's estimate that atomic numbers were one-half of atomic weights.

It was a happy shop in those days. Rutherford was everywhere, encouraging, goading, and every once in a while demonstrating his explosive temper, then regretting it. He blew up when Geiger told him that lab work had to be shut down because radiation emanations from Rutherford's own bench were contaminating every other experiment. Rutherford stormed out, returning in ten minutes to ask Geiger to come with him. They climbed into the Wolseley-Siddeley and drove through the countryside talking physics. "Nothing was so refreshing and inspiring as to spend an hour in this way, alone with Rutherford," Geiger wrote years later. "In spite of the minor provocation, I would be loth to part with the memory of such a day, spent in fellowship with a master-mind."[6]

Moseley was traveling with his mother in Australia in August 1914, when World War I began. He booked passage home immediately and in October was commissioned a lieutenant in the Royal Engineers. He was a signal officer of the Thirty-Eighth Brigade, Thirteenth Infantry Division, which landed on the Gallipoli Peninsula of Turkey on August 6, 1915. At dawn on August 10, the Turks, more than thirty thousand of them, attacked from an 850-foot hill above the British forces. The poet John Masefield, who survived the battle, described it: "They came on in a monstrous mass, packed shoulder to shoulder, in some places eight deep, in others three or four deep . . . The Turks finally got in and among our men with a weight which bore all before it, and what followed was a long succession of British rallies to a tussle body to body with knives and stones and teeth in the ruined cornfields."[7]

Captain Harry Moseley was killed there that day, shot through the head by a Turkish bullet. He was twenty-seven years old. Exactly two weeks before, he had written a will leaving everything he had, twenty-seven hundred pounds sterling, "to be applied to the ·furtherance of experimental research in Pathology, Physics, Physiology, Chemistry or other branches of science, but not in pure mathematics, astronomy or any branch of science which aims at merely describing, cataloguing or systematizing."[8]

"It is a national tragedy," wrote Rutherford. "Our regret for the untimely death of Moseley is all the more poignant because we recognize that his services would have been far more useful to his country in numerous fields of scientific enquiry rendered necessary by the war."[9] An American physicist, Robert Millikan wrote later of Moseley's death, "Had the European War no other result than the snuffing out of this young life, that alone would make it one of the most hideous and most irreparable crimes in history."[10] Rutherford had secretly tried to have Moseley returned home before he reached the bloody combat of Gallipoli. The Ministry of War, it seems, agreed that Moseley would be a more valuable asset at home than on the battlefield, and was ready to recall the young officer. But it was too late. 3

War indeed needs science, and science has traditionally thrived on the high proportions of public money diverted to research and development in time of war. But war is a time of choosing for men of science, often ignored in peacetime but coveted in wartime. Hans Geiger, conscripted as a German artillery officer, had already been wounded. E. N. Andrade was in the British artillery. James Chadwick, one of Rutherford's team, was a British citizen interned in a German prison camp because he had a fellowship in Berlin; he happened to be in the wrong country at the wrong time.

Along with four thousand other "enemy aliens," Chadwick was taken to a racecourse two miles west of Berlin and held there for three years. His "home," shared with five others, was a ten-foot-square stall designed for two horses. Ignoring the cold and the smells of urine and dung, Chadwick set up a makeshift laboratory to conduct radioactivity experiments. Geiger was able to bring him some equipment, and he used a German toothpaste containing traces of thorium—radioactive material was still being sold as a cosmetic product promising whiter teeth and softer skin—as his source. He improvised a torch to blow glass by igniting rancid butter blown through an asbestos cone.

The master glassblower Otto Baumbach, an outspoken German chauvinist, was jailed in England. Otto Hahn, the German chemist who had worked with Rutherford at McGill, led a team in Berlin developing poison gas.* Arthur Schuster,

---

* Hahn's work included testing gases and supervising their use on battle-fields. Like others who worked on poison gas—and later on the atomic bomb—he justified the work by saying they believed it would end the war sooner. "The wind was favorable and we discharged a very poisonous gas, a mixture of chlorine and phosgene against enemy lines," he wrote after one battle in 1915. "Not a single shot was fired. The attack was a complete success" (Richard Rhodes, *The Making of the Atomic Bomb* [New York: Simon & Schuster, 1986], 93). Hahn's immediate supervisor in those days was Fritz Haber, whose wife, Clara Immerwahr Haber, was a chemist at the University of Breslau. She begged Haber to give up the work, and when he refused, rushing off to supervise an attack, she killed herself. During World War II, Hahn stayed in Germany. In 1944, he was awarded the Nobel Prize in Chemistry as the discoverer of nuclear fission—an award many felt should have been shared with his longtime collaborator, Lise Meitner, who

Rutherford's predecessor and patron at Manchester, was being attacked in newspapers because of his German ancestry on the same day he learned that his son, a British soldier, had been wounded at Gallipoli.

Rutherford, too, was in Australia in the summer of 1914 when the war broke out in Europe. In fact, both he and Moseley were at an Australian Royal Society meeting, at which the older man had made a point of introducing his protégé and publicly praising both the importance of the younger man's work and his brilliant potential. But as Moseley rushed back to England to enlist, Rutherford, traveling with his wife, continued on a long-planned journey through Australia, home to New Zealand, and then on to the United States and Canada, lecturing and collecting awards, medals, and honorary degrees. In New Zealand, as Lord and Lady Rutherford spent time with their families, there were almost daily tributes to the islands' favorite son: lectures, dinners, even parades with students pulling his car through the streets of Christchurch. Most important, Rutherford met with scientists everywhere, including former students, discussing their work and his own. Then he detoured to North America, visiting his old friends at McGill and at Columbia and other American universities.

The war had been going on for almost six months before Lord Rutherford returned to England on January 15, 1915. One of the first things he did was to accept appointment to

---

had fled Germany in 1938 because of her Jewish ancestry. Hahn could not travel to accept the prize, because he was by then, in effect, a prisoner in Germany. Although his work was critical to the development of atomic weapons, Hahn had refused to work on the German atom bomb project, and he spent his later years as an antibomb writer and activist.

the Royal Navy's Anti-Submarine Division, a controversial and largely thankless position because, like many British institutions, the Royal Navy stoutly resisted any input from outsiders, no matter how talented or famous—which was one reason the Germans were able to convert the science of gas and of submarines into potent weaponry long before the British. Those who knew Rutherford well understood that he loved Great Britain and its ways, but his true nation was science. He saw scientists and scientific communities as different from other citizens—more important than politics and parliaments. More than once he was criticized for a lack of "public spirit."

There was some truth in that, particularly if "public spirit" was another name for nationalism. Late in the war, Rutherford ran a series of experiments at Manchester that climaxed with the breaking off or chipping of part of a nitrogen nucleus. Staying in the lab, he missed an Anti-Submarine Division meeting, telegraphing this excuse: "If, as I have reason to believe, I have disintegrated the nucleus of the atom, this is of greater significance than the war."[11]

"Disintegrated" may have been too strong a word, but Rutherford had succeeded in knocking off part of the nucleus of nitrogen atoms; it was the first observation of nuclear disintegration brought about by artificial means. And he did it with a small brass tube, two glass valves, and brass extensions. The apparatus, on display now at the Cavendish Laboratory, is less than a foot long; the cylinder is about four inches high. Like the exhibits at McGill, the Cavendish apparatus is a thing of beauty, displayed among other little boxes and cylinders, tiny symphonies of brass and glass with the joints sealed by the old red sealing wax.

The nitrogen experiments had been started by Ernest Marsden working under Rutherford's supervision. But in May 1915, Marsden, one of the first members of Rutherford's Manchester tribe, left to take the physics chair at the University of New Zealand. The appointment, of course, was on Rutherford's recommendation—he was becoming J. J. Thomson's successor as the high recommender of international physics—and he had provided Marsden with an invaluable going-away gift: a small supply of radium from the Royal Society for use in new radioactivity experiments. Then, like most of the rest of Rutherford's young men, Marsden went to war, serving as an officer in the New Zealand Expeditionary Force and winning a Military Cross in France.

Marsden was already on his way to New Zealand when Rutherford took a closer look at his research. At Rutherford's suggestion, Marsden had performed a series of experiments bombarding hydrogen atoms with alpha particles and finding that some of the atoms were shooting far beyond the range of the particles themselves. Using variations of scattering experiments, Marsden saw that what he called "H-particles," speedy hydrogen atoms hit by alpha particles, were banging brightly against scintillation screens. In two papers that he wrote hastily before leaving England, Marsden suggested that the H-particles were apparently coming from within the radioactive sources that he used; in other words, there might be a fourth ray emitted, after alpha, beta, and gamma rays.

That conclusion seemed wrong to Rutherford, and he wrote to Marsden, already in military training, asking if Marsden would "mind" if Rutherford himself continued the series of experiments. Of course not, came the answer. Then, over

The house where Ernest Rutherford was born in 1871, built by his father, James, on the frontier settlement of Spring Grove (later renamed "Brightwater") on South Island, New Zealand. CREDIT: ALEXANDER TURNBULL LIBRARY, WELLINGTON, NEW ZEALAND

James Rutherford's flax mill at Pungarehu on North Island, New Zealand, where Ernest was digging potatoes when he learned he had won a scholarship to Cambridge in England. CREDIT: ALEXANDER TURNBULL LIBRARY, WELLINGTON, NEW ZEALAND

James and Martha Rutherford in a formal photograph. CREDIT: ALEXANDER TURNBULL LIBRARY, WELLINGTON, NEW ZEALAND

The Rutherford family at Havelock in the early 1880s. From left to right: Alice Rutherford, Mary Thompson (cousin), Arthur Rutherford, Ernest, James Rutherford, Eve Rutherford, Nell Rutherford, Ethel Rutherford, George Rutherford, Herbert Rutherford, Flo Rutherford, Martha Rutherford, Charles Rutherford, Jim Rutherford. Photo taken by William Collie before 1886. CREDIT: ALEXANDER TURNBULL LIBRARY, WELLINGTON, NEW ZEALAND

The gate of the Cavendish Laboratory on Free School Lane in Cambridge. CREDIT: CAVENDISH LABORATORY, UNIVERSITY OF CAMBRIDGE

Rutherford, far left in the back row, among classmates at Cavendish in 1895. CREDIT: ALEXANDER TURNBULL LIBRARY, WELLINGTON, NEW ZEALAND

The physics faculty at McGill University, Montreal. Rutherford is seated on the left. His official biographer, Arthur Eve, is standing in the middle, sixth from the left.

Rutherford at Manchester, holding the simple apparatus he used to bombard nitrogen atoms with alpha particles and, for the first time, disintegrate a nucleus—an experiment that created nuclear physics and would lead to the development of the atomic bomb.

The thirty-four-year-old professor of physics at McGill, soon to win a Nobel Prize, was photographed formally in his laboratory in 1905. In a story told many times, Rutherford's white cuffs were borrowed from a student, Otto Hahn, when the photographer complained that Rutherford did not look old enough or dignified enough to be such an eminent fellow.

Rutherford in his thirties. He looked, said one student, like a shopkeeper in backcountry Australia. Credit: AIP Emilio Segrè Visual Archives (gift of Otto Hahn and Lawrence Badash)

Rutherford (right) and Hans Geiger posing for a photograph in the basement physics laboratory of the University of Manchester in England. CREDIT: ALEXANDER TURNBULL LIBRARY, WELLINGTON, NEW ZEALAND

Rutherford (seated fourth from the right) with his staff at Manchester. Marsden is second from the right in the back row. Geiger is second from the right in the middle row. CREDIT: ALEXANDER TURNBULL LIBRARY, WELLINGTON, NEW ZEALAND

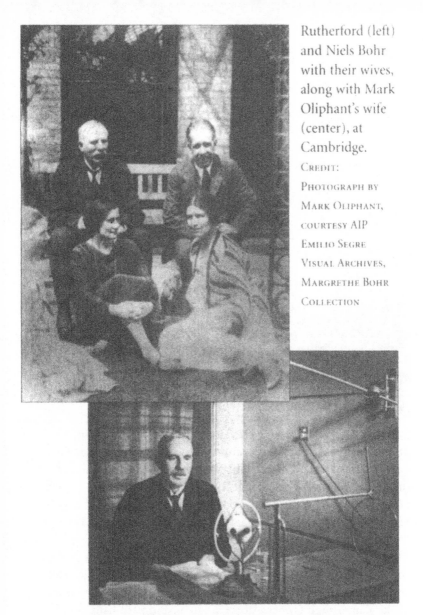

Rutherford (left) and Niels Bohr with their wives, along with Mark Oliphant's wife (center), at Cambridge. CREDIT: PHOTOGRAPH BY MARK OLIPHANT, COURTESY AIP EMILIO SEGRE VISUAL ARCHIVES, MARGRETHE BOHR COLLECTION

Rutherford, shown here at a BBC microphone in the early 1930s, became a popular radio speaker on science. His colleagues joked that he didn't need a microphone to reach such places as the United States and Australia. CREDIT: ALEXANDER TURNBULL LIBRARY, WELLINGTON, NEW ZEALAND

The first Solvay conference, sponsored in 1911 by a Belgian physicist, Ernest Solvay, to "evaluate some theories of molecules and motion" attracted many of the finest scientists and thinkers of the century. Seated, left to right: Walter Hermann Nernst, Marcel-Louis Brillouin, Solvay, Hans Lorentz, Otto Heinrich Warburg, Jean Perrin, Wilhelm Wien, Marie Curie, and Jules-Henri Poincaré. Standing, left to right: Robert Goldschmidt, Max Planck, Heinrich Rubens, Arnold Sommerfeld, Frederick Lindemann, Maurice de Broglie, Martin Knudsen, Friedrich Hasenöhrl, Georges Hostelet and Edouard Herzen (Solvay's assistants), James Jeans, Rutherford, Heike Kamerlingh Onnes, Albert Einstein, and Paul Langevin.

A 1930s gathering of friends in Münster, Germany. From left to right: Seated are James Chadwick, Hans Geiger, Rutherford, and two unidentified men. Standing are Georg Karl von Hevesy, Mrs. Elisabeth Geiger, Lise Meitner, and Otto Hahn. CREDIT: ALEXANDER TURNBULL LIBRARY, WELLINGTON, NEW ZEALAND

J. J. Thomson and Rutherford (right) outside Cavendish in the late 1930s. Thomson, the former director, often said that his greatest pleasure was in turning over the world's greatest laboratory to the best student he ever had. CREDIT: CAVENDISH LABORATORY, UNIVERSITY OF CAMBRIDGE

Rutherford's laboratories looked the same from boyhood to old age. He said he felt sorry for men who did not have one, and this is what his looked like at Cavendish in 1933. CREDIT: CAVENDISH LABORATORY, UNIVERSITY OF CAMBRIDGE

"Talk Softly Please," a famous photograph of Rutherford (right) ignoring the sign at Cavendish Laboratory, which was put up because his booming voice and singing of "Onward, Christian Soldiers" shook equipment and men as well. CREDIT: PHOTOGRAPH BY C. E. WYNN-WILLIAMS, COURTESY AIP EMILIO SEGRE VISUAL ARCHIVES

"The Prof" and some of his "boys" and men in the courtyard of the Cavendish. From left to right: Patrick Blackett, Pyotr Kapitsa, Paul Langevin, Rutherford, and C. T. R. Wilson. Langevin, Rutherford, and Wilson had been classmates at Cavendish.

Cockcroft working at the desk that controlled the particle-accelerating mechanism, essentially the cannon that split the atom. CREDIT: THE ILLUSTRATED LONDON NEWS PICTURE LIBRARY

The main photograph of the *Illustrated London News* story reporting that Rutherford's boys, Ernest Walton and John Cockcroft, had won the worldwide race to split the atom. The original caption read, "A is the transformer, B a condenser, C the four-stage glass rectifier tower, D the acceleration tube, E the observation hut (with black curtain drawn) and F the spark gap spheres. The two unidentified towers in the background are another condenser set and the transformer for the proton supply apparatus (far right)." CREDIT: THE ILLUSTRATED LONDON NEWS PICTURE LIBRARY

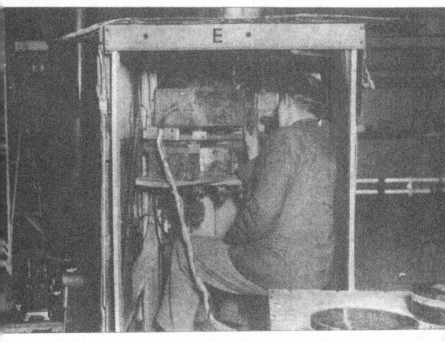

Walton in the "hut," which was actually a tea crate—exactly the same kind of crate thrown off British ships by Americans during the Boston Tea Party. It was tough work for Rutherford, stouter in old age, to get into the thing without help. CREDIT: THE ILLUSTRATED LONDON NEWS PICTURE LIBRARY

Walton, Rutherford, and Cockcroft posing for news photographers on May 2, 1932. CREDIT: THE ILLUSTRATED LONDON NEWS PICTURE LIBRARY

The 1933 Solvay conference. Rutherford is seated sixth from the right. Also seated are Marie Curie (fifth from the left); her daughter, Irène Joliot-Curie (second from the left); Niels Bohr to Joliot-Curie's left; Lise Meitner (second from the right); and James Chadwick (far right). Among those standing are Werner Heisenberg (fourth from the left), John Cockcroft (sixth from the left), and Ernest Lawrence (second from the left). CREDIT: © CORBIS

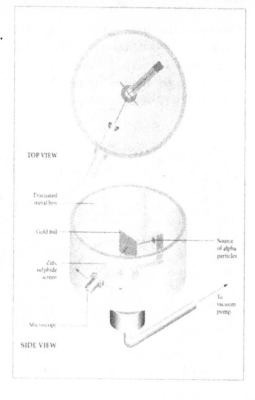

TOP VIEW

Evacuated metal box

Gold foil

Zinc sulphide screen

Source of alpha particles

To vacuum pump

Microscope

SIDE VIEW

More than 10,000 people filled Royal Albert Hall on October 3, 1933, at a meeting called and hosted by Rutherford to raise funds for scientists being driven from Germany by the government of Adolf Hitler. Rutherford is seated just behind the main speaker of the night, Albert Einstein. Within a week, Einstein sailed for the United States and Princeton University's Institute for Advanced Studies. CREDIT: © BETTMANN/CORBIS

(facing page) The scattering experiment. In a vacuum chamber, positively charged alpha particles are fired through a slit from a radioactive source. Most pass straight through the gold foil, producing a sharp, straight line on the screen painted with zinc sulfide in front of the microscope. Some, however, are deflected by the positive charge of gold nuclei, blurring the edges of the zinc sulfide line and making visible hits just outside the line. Others came back at angles of 90 degrees and more. One in 8,000 particles came almost straight back at close to 180 degrees after a near head-on collision with a nucleus. CREDIT: JOHN MCAUSLAND

The experiment re-created at Stevens Institute of Technology in 2005. From left to right: Damien Marianucci, Richard Reeves, Kurt Becker, and Frank Corvino. Credit: Jim Cummins

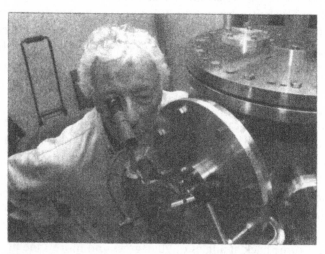

The author, in the Hans Geiger position at the viewing microscope. Credit: Jim Cummins

time, conducting hundreds of test runs by himself, Rutherford, who was at the same time taking days and weeks off for his Royal Navy meetings, most of them frustrating wastes of time, came to his own conclusion: the alpha particles were chipping away at nitrogen nuclei, producing positively charged hydrogen atoms. If that was so, the process was transforming the nitrogen atoms into oxygen atoms. So it could be said that Marsden had succeeded as early as 1914 in "splitting" an atom, turning it into different elements: nitrogen into oxygen. But he did not know that. By 1917, however, Rutherford did.

In 1918, his military service at an end, Marsden came back to Manchester, telling Rutherford that he would be sailing for New Zealand again in a month to take up his delayed professorship, and . . . He never got to finish the sentence. "You'll stay with us, of course," said the professor. "I want to show you something."[12]

"Remember the experiment you were working on?" Of course. Marsden saw that Rutherford had replaced his glass apparatus with the small brass cylinder because metal is less susceptible to radioactive contamination than is glass. Rutherford went over the tests he had run while Marsden was at war, saying that he had confirmed Marsden's conclusion that the little missiles he discovered were speeding H-particles, but that they did not, as Marsden had thought, come from inside a radioactive source.

Rutherford had sealed one end of the cylinder with silver foil thick enough to stop alpha particles but not the H-particles. Over the next few days, Rutherford filled the cylinders with air and its components. In a dry-air test, there were some scintillations as the H-particles shot through the silver

foil to hit the zinc sulfide target. Then he tried oxygen—
nothing. Carbon dioxide—nothing. Water vapor—nothing.

"Now watch this," Rutherford said to Marsden. Turning the
valves on the glass tubes coming out of the top of the cylinder,
Rutherford filled the cylinder with nitrogen. The scintillation
target lit up as it never had before. Rutherford told Marsden
that he had used magnetic fields to verify that the particles
hitting the screen were hydrogen nuclei. In other words, alpha
particles (helium nuclei) were "hitting" nitrogen nuclei and
driving out a positively charged hydrogen atom. Rutherford
gave that particle a new name: the "proton." The process had,
in effect, changed one element (nitrogen) into others (hydrogen and oxygen). Scientists could artificially disintegrate matter as radioactivity did naturally. It was alchemy.

Like most successful experiments, this result led to another
question: what happened to the rest of the chipped nitrogen
atom? It would take years and the end of World War I to figure out the answer.

Because of Britain's Official Secrets Act, it would also be
years before Rutherford's contributions to the country's war
effort were known to the public. Science, engineering, and
government came together in World War I in productive and
deadly combinations: submarines, poison gas, machine guns,
aerial bombing. One of the secret heroes of Britain's effort
was Rutherford's friend, the chemist Chaim Weizmann, who
spent two years developing methods for the mass production
of acetone, a key ingredient in artillery explosives like
cordite—and who used that valuable service and new friend-
ships with well-placed officials in the British government to
gain critical support for his own cause, Zionism, the creation
of a Jewish state in Palestine.

Rutherford's role in the war was not widely known until decades later. Some secret government records, particularly those critical of the effects of the smug snobbery of the Royal Navy, became public in the 1960s. Then, in 1983, Rutherford's antisubmarine research was the subject of more than forty-five pages of David Wilson's book *Rutherford, Simple Genius*. Using official records and correspondence for the first time, Wilson, writing of the last three months of 1915, said this: "In three reports [to the Admiralty] he drew up the 'map' of underwater warfare which has remained unchanged to the present day. What Rutherford said about submarine-hunting in 1915 remains true in the 1980s."[13]

Perhaps even more important, the most prominent scientist in the realm, it was revealed, was an architect of the cooperation of science and government, which paid real dividends in World War II when the British developed "radar" (which stands for "radio detecting and ranging") and exotic code-breaking machines. But Rutherford's war work, which included tearing up his own labs to build water tanks for testing listening devices, was no easy job from the beginning. The story of the Royal Navy's initial disinterest in antisubmarine warfare was not unlike General Billy Mitchell's inability to persuade the U.S. army and navy that future wars might well be decided in the air.

Even as German U-boats were crippling Britain's dominant power at sea, the Royal Navy tried to cripple scientific research into underwater warfare. W. H. Bragg, the Australian whom Rutherford had first met on his way to England back in 1895, was a member of the committee and told a story of being handed an "urgent" Admiralty order as he began a meeting with French scientists; it read, "On no account show them anything."[14]

It was doubtful that anyone with less prestige and power than Rutherford could have forced admirals to deal seriously with scientific and technological progress. He began with his usual enthusiasm. Records and letters picture him bobbing in a small trawler on Scotland's cold and choppy Firth of Forth. The little boat was about the only thing the Admiralty would give the antisubmarine team. The navy would provide no larger vessel, and actual submarines were out of the question. So Rutherford was told, as he spent long and windy days in the open. "I spent three days hard at work on a converted 'trawler' and I expect to go up again shortly," he wrote to his mother. ". . . We heard last night of the submarining of a troop ship in the Aegean with a loss of 1,000 lives."[15]

Part of Rutherford's work in Scotland was holding tightly the legs of Sir Richard Paget, whose head was underwater listening for distinctive sounds. A human speech expert at Cambridge, Paget was believed to have a perfect sense of pitch, and Rutherford was trying to calibrate the reception of underwater microphones. On the scientific level, Rutherford had the basic insights that made possible the intensive and expensive research on passive listening for sounds emitted by submarines, and then later an active technology using sound aimed to bounce off objects underwater. He was, as always, the hands-on experimenter, writing in one report, "We have found in testing diaphragms it is very important to avoid grease on them and make certain that bubbles of air are not attached to the diaphragm. We had been noting variations in resonance points and finally traced them to Powell's industry in carefully greasing the diaphragms to prevent rust! A clean surface is essential if results are to be repeated. Grease layer usually lowers the frequency of resonance while bubbles raise it. Keep your eyes open on this point."[16]

That research led to the beginning of the development of ASDIC (Anti-Submarine Detection Investigation Committee) devices; ASDIC was renamed "sonar" (for "sound navigation and ranging") by Americans in World War II. This was the sound wave–bouncing technology that was refined to locate the positions of silent submarines deep under the surface of the sea. Papers and testimonies of the period credit Rutherford as the inventor of ASDIC, which he demonstrated on a seven-week trip to the United States after the Americans entered the war in 1917.   3   Sonar

Rutherford found the Americans a year or two behind the British in antisubmarine technology but far more open to research and cooperation than was the Royal Navy. "We found the American authorities were very conscious of the great potential value of science in warfare," he reported seven days after his return to England. "But it was soon apparent that they were much handicapped owing to lack of knowledge of the practical conditions and complete want of information regarding the scientific work already carried out in England and France . . . We found the American authorities most anxious to receive and act upon the advice we were able to give . . . We do not wish to imply that the American authorities are lacking either in energy or initiative. Very much the reverse is the case."[17]

One of the American authorities whom Rutherford met was Thomas Edison, the eighty-year-old chairman of the U.S. Naval Consulting Board. Edison was well past his prime and almost deaf, but Rutherford loved him as a kindred soul, writing home, "I was received very well by the old man, who was as enthusiastic as a schoolboy over his ideas."[18] And then he stopped by to collect honorary degrees from both Yale and Harvard.

Back home, frustrated by the Navy, Rutherford began to concentrate again on his own atom-smashing experiments. Though in one six-month stretch of submarine research he had not taken a single day off, his war work was forgotten, hidden in old letters and records until Wilson came along. The accomplishment of antisubmarine technology was generally credited to the French, specifically to Rutherford's old friend from across the hall at Cambridge, Paul Langevin. Certainly both men were working along the same acoustic lines during the war—and coming up with the same findings without any apparent communication between them. (Langevin was provided with a submarine to use in experiments in the Bay of Toulon; Rutherford and other British scientists publicly were not allowed to experiment with the real thing until the very end of the war.) When other British scientists publicly asserted that Rutherford deserved at least equal credit, he silenced them, saying privately, "If Langevin says the idea was his, then the idea was Langevin's."[19]

His war was long over. As early as December of 1917, he had written to Bohr: "I have got, I think, results that will ultimately have great importance. I am detecting and counting the lighter atoms set in motion by alpha particles and the results, I think, throw a great deal of light on the character and distribution of forces near the nucleus." And then he wrote a sentence that would make scientific history: "I am trying to break up the atom by this method . . . Best wishes for a happy Christmas."[20] Rutherford waited until 1919 to publish four cautious papers in *Philosophical Maga-*

zine on his wartime scattering experiments under the title "Collision of Alpha Particles with Light Atoms." He reported, "From the results so far obtained it is difficult to avoid the conclusion that the long-range atoms arising from collision of alpha particles with nitrogen are not nitrogen atoms, but probably atoms of hydrogen . . . If this be the case we must conclude that the nitrogen atom is disintegrated under the intense forces developed in a close collision with a swift alpha-particle and that the hydrogen atom which is liberated formed a constituent part of the nitrogen atom."[21]

In other words, nitrogen went into one end of the little brass cylinder, and hydrogen and oxygen came out the other. It happened only about once each three hundred thousand times as alpha particles were fired into the nitrogen-filled cylinder. But it happened, and it was treated as major news in British newspapers. In the United States, *The New York Times* headlined the story

WAY TO TRANSMUTE ELEMENTS IS FOUND
*Dream of Scientists for a Thousand Years Achieved by
Dr. Rutherford*[22]

For some reason it took almost two years for American newspapers to catch up with scientific journals. In New York, *The Times* published two stories on the 1919 experiments— on December 14, 1921, and January 8, 1922—both based on a speech by O. W. Richardson of Kings College, London, to the British Association in New York and follow-up interviews with American physicists. The newspaper quoted James Kendall of Columbia as saying, "This is certainly the transmutation of elements, but it is done as an infinitely small scale and is important at present only to the scientific man."[23]

Richardson was then quoted:

A careful and critical examination of [Rutherford's] results shows . . . the artificial transmutation of chemical elements is thus an established fact. The natural transmutation has, of course, been familiar for some years to students of radioactivity. The philosopher's stone, one of the alleged chimeras of the medieval alchemists, is thus within our reach. But this is only part of the story. Now we know from the heat liberated in radioactive disintegration that the amount of energy stored in the nuclei is of a higher order of magnitude, some millions of times greater, in fact, than that generated by any chemical reaction such as the combustion of coal . . . If these effects can be sufficiently intensified there appear to be two possibilities. Either they will prove uncontrollable, which would presumably mean the end of all things, or they will not. If they can be both intensified and controlled, then we shall have at our disposal an almost illimitable supply of power which will entirely transcend anything hitherto known . . . It may be that we are at the beginning of a new age, which will be referred to as the age of sub-atomic power. We cannot say; time alone will tell.[24]

"Intensification" was also the conclusion of Rutherford's 1919 paper: "If alpha particles—or similar projectiles—of greater energy were available for future experiment, we might expect to break down the nucleus structure of many of the lighter atoms."[25]

A call for "greater energy," more powerful projectiles, coming from the greatest of tabletop experimenters could be taken as a strong, if indirect, suggestion that it might be time for bigger science. The obvious next step in probing the subatomic world was machines capable of speeding up nuclear particles—to continue to learn about the atom by more intense demolition of nuclei. That is, the time had come to split the atom on an industrial scale.

That would take years, however, because the war—a terrible war—had already demolished or diminished not atoms, but the many scientific institutions studying them in many countries, killing off many of the scientists and engineers capable of working on the causes and effects of atomic power. One of the places hardest hit was Rutherford's old home, Cavendish. It was still run by J. J. Thomson, who was sixty-one years old at the end of the war. But by 1918 the laboratories were being used as an officers' billet. Material and men had been steadily stripped away from it, leaving it with only a thousand pounds sterling on hand. On March 5, 1918, Thomson was appointed master of Trinity College, though he fully intended to keep his appointment as the Cavendish professor. But that was not to be.

Exactly a year later, in March 1919, Thomson was persuaded to leave as the director of Cavendish and accept election to a new research professorship. The obvious choice to become the new director was Rutherford. But, at forty-nine, he was already the highest-paid professor in England, and he was wary, too, of any kind of shared responsibility with his old mentor. The negotiations took a while; at issue, after all, was one of the most prestigious posts in a country that revered prestige and title. At one point in the negotiations about budgets and such, Cambridge's best-known chemistry

3

professor, W. J. Pope, wrote to Rutherford and to Thomson urging them to work this out for "the greater Glory of God and the advancement of Physics and Chemistry."[26] They did.

Whatever his feelings were about being moved out, Thomson arranged to increase Rutherford's pay to match his Manchester salary by granting Rutherford a fellowship at Trinity College—the fellowship that Rutherford had been denied more than twenty years before—and then writing Rutherford a gracious letter, saying, "There is very keen hope that you may see your way to come to Cambridge. Nothing would give me more pleasure as to have my successor be my most distinguished pupil."[27]

utherford finished out the 1919 school year at Manchester, packed up the apparatus to continue the proton experiments, and took on the most prestigious position in British physics: Cavendish. It was a triumphant homecoming for the boy from New Zealand who had first walked through the soot-colored arch on Free School Lane as a colonial scholar twenty-four years before. He was greeted there at a festive dinner by students singing, arms crossed, in the wacky way of their old tribe:

> We've a professor
> A jolly smart professor,
> Who's director of the lab in Free School Lane,
> He's quite an acquisition
> To the cause of erudition, . . .
> When he first did arrive here,
> He made everything alive here,
> For, said he, the place will never do at all;

I'll make it nice and tidy,
And I'll hire a Cambridge "lidy"
Just to sweep the cobwebs from the wall

And what he's been achieving
Would be almost past believing
If he weren't quite a marvel among men . . .
What's in an atom,
The innermost substratum?
That's the problem he's working on today.
He lately did discover
How to shoot them down like plover,
And the poor little things can't get away.
He uses as munitions
On his hunting expeditions
Alpha particles which out of Radium spring.
It's really most surprising
And it needed some devising,
How to shoot down an atom on the wing.[1]

The dust and cobwebs did look as if they had been growing there since Cavendish opened in 1871. But buildings and cleanup were at best a part-time fashion for Rutherford. Marcus Oliphant, the young Australian who arrived in 1927, described the director's office: "I entered a small office littered with papers, the desk cluttered in a manner which I had been taught at school indicated an untidy and inefficient mind . . . It was raining and drops of rain ran reluctantly down the grime-covered glass of the uncurtained window . . . I was received genially by a large, rather florid man with thinning fair hair and a large moustache, who reminded me forcibly of the keeper of the general store and post office in a little village in the hills behind Adelaide . . . Rutherford made me feel welcome and at ease at once. He spluttered a little as he talked, from time to time holding a match to a pipe which produced smoke and ash like a volcano."[2]

Cavendish was, and had been for a very long time, a very old-fashioned place. The greatest change since Rutherford had first arrived there in 1895 was the number of students. There were six hundred now, almost half of them research students needing laboratory room and time, overcrowding the place. The numbers were driven up by veterans (two hundred of them Americans), many trying to catch up with their lives, returning to school after years of war. Rutherford immediately rejected the suggestion that the number could be reduced by refusing admission to women—not too tough a decision for a man who had worked with Marie Curie and her daughter, Irène Joliot-Curie, and with the great Austrian physicist Lise Meitner.* On December 6, 1920, Rutherford and William J. Pope, the chemistry professor, cosigned a letter to *The Times* of London saying,

> For our part, we welcome the presence of women in our laboratories on the ground that residence in this University is intended to fit the rising generation to take its proper place in the outside world, where, to an ever increasing

---

* Rutherford had read of Meitner's work for years, but when he met her for the first time he said, "My God, I always assumed you were a man" (Diana Preston, *Before the Fallout: From Marie Curie to Hiroshima* [New York: Walker, 2005], 49). In a poll of physicists in the 1990s, there was a question about which practitioner who had never won one most deserved a Nobel Prize; the majority of respondents named Lise Meitner.

extent, men and women are being called upon to work harmoniously side by side in every department of human affairs. For better or for worse, women are often endowed with such a degree of intelligence as enables them to contribute substantially to progress in the various branches of learning; at the present stage in the world's affairs we can afford less than ever before to neglect the training and cultivation of all the young intelligence available. For this reason, no less than for those of elementary justice and of expediency, we consider that women should be admitted to degrees and to representation in our University. 3

The new director also was still attached to the old Cavendish. He continued the six-o'clock lab locking that he had learned from Thomson, sending demonstrators, researchers, and students home each evening. He climbed the old stairways and walked the winding hallways and courtyards almost every day to check in on every room and table, offering encouragement and advice, demanding results: "When on earth will you get results? . . . Will you potter around for a long time to no purpose? . . . I want you to give me results, results, not your chatter."[3] A story, perhaps apocryphal, was told for years about a bricklayer working on a new wall to divide laboratories who walked off the job saying, "I'm not going to put up with a rude and gruff old gentleman who wanders around and around asking why I'm not giving results."[4]

At afternoon tea, the new director listened to the problems and weathered the brainstorms of his ambitious charges. On Sundays, his senior men and chosen students would be invited to the house, Newnham Cottage, where Rutherford's wife, Mary, who became more severe as the years went by, still refused to allow any alcohol in her home. She sent everyone away at an appointed hour, shaking hands and saying good-bye with finality or, in good weather, offering tours of her gardens that ended at a back gate that suddenly led company out into the street.

Rutherford also kept Cavendish nineteenth-century cheap. Like the few and eminent directors before him—Thomson, John Strutt (Baron Rayleigh), and the founder, James Clerk Maxwell—he believed that shabby poverty produced more innovative results. One story that everyone loved to tell was of a young researcher who approached the director on one of his black days and asked for a few shillings for a new spring balance scale. "Why must you always have money?" was the answer. "Why can't you improvise? Why don't you hang beakers on the ends of the wires and load them by pouring in water?" The young man replied that it would cost more money to buy beakers that could hold twenty-five pounds of water—and they would inevitably end up as piles of wet glass on the lab floor and new ones would be needed. Rutherford looked at him for a moment, then laughed, said yes to a new scale, and began a verse of "Onward, Christian Soldiers."[5]

Rutherford was as cheap as any of his Scottish forebears, and he did not like to ask for money. Though he could have used his own fame, political clout, and foreign connections to raise large amounts of research funding, he refused to consider it. In general, he was willing to do genteel begging only to get more radioactive material for his own research and that of others, beginning with James Chadwick, his former Manchester student, who came back after his wartime internment at the age of twenty-seven depressed, malnourished, and broke. Chadwick was taken in by the Rutherfords on Christmas of 1918, immediately made a part-time instructor at Manchester, and then assigned to the post of deputy director at Cavendish. 3

Only years after Rutherford's death, in fact, did Chadwick, who had been in charge of the budget (and budget problems) at Cavendish, learn that the family that owned the Orient Steam Navigation Company had offered to supply Rutherford with any funds he needed, if he asked. But he never did. Rutherford never told Chadwick about the offer. He never told his wife. Looking back, Chadwick thought he knew why:

> I believe he was essentially a very modest man. I think he was always surprised at what he had done . . . During one of our talks, he suddenly said—I say suddenly because it had little connection with our talk on writing papers for publication—"You know, Chadwick, I sometimes look back on what I have done and wonder how on earth I managed to do it all" . . . I knew him fairly well, because it wasn't only in the laboratory that I knew him. I knew him at home . . . I knew him in off-guarded moments when he was not being the great physicist . . . That, I think was modesty. It was also a reluctance to be responsible to somebody outside the laboratory. He did not feel he could justify spending so much money on himself and his research students. And this in spite of his unique position in the scientific world, his extraordinary achievements in the past.[6]

Other old students and old friends surrounded Rutherford at Cavendish, including Charles G. Darwin. Then there was C. T. R. Wilson, inventor of the cloud chamber—a vessel of moist air producing droplets that attached themselves to particles, leaving a visible trail—which, after 1911, had allowed Rutherford and others to follow (and later photograph) the tracks of the alpha particles that they could never actually see. Perhaps the greatest symbol of the old days kept new was Frederik Lincoln, the lab steward, a tyrant with a junk-yard mentality and the heart of an equipment cannibal. When one research student, Edward Bullard, worked up his courage to tell Lincoln that he needed a one-inch steel pipe for an experiment, he was surprised when the steward replied that that was no problem. Lincoln handed Bullard a hacksaw and pointed to a rusting old bicycle that had been leaning against a courtyard wall for as long as anyone could remember. Actually smiling, Lincoln said, "Just cut off some of the handlebar."[7]

Above all the scrambling for equipment and for results was that big voice moving Cavendish forward slowly and science forward more rapidly than most anyone before him. A large sign appeared in the laboratories that said, "TALK SOFTLY PLEASE"—a not very successful warning that Rutherford's personal sound waves were a danger to experiments of any delicacy. Told that Rutherford was scheduled to make a radio broadcast from Cambridge, England, to Cambridge, Massachusetts, the home of Harvard, one of his men said, "Why use radio?"[8] As he grew older and larger and a bit shaky too, Rutherford's clumsiness was becoming more of a threat to equipment and experiments—to say nothing of nerves. One researcher offered the irreverent opinion that tweezers must be hidden from the boss because he would surely drop them under a table or desk, never to be seen again.

In 1920, Rutherford was asked to give the Bakerian Lecture to the Royal Society for the second time—the first had been in 1904—and after the usual review of recent progress in physics of all kinds, he began what sounded like speculation about the future. But it was more than that, because at least three of his ideas proved out. He wondered about "heavy hydrogen" with a nucleus mass of two units rather than one and a single electron. He talked of a lighter helium isotope with a mass of three units and a charge of two units. Then, most amazingly, he made this prediction:

> It may be possible for an electron to combine much more closely with the hydrogen nucleus, forming a kind of neutral doublet. Such an atom would have very novel properties. Its external field would be practically zero, except very close to the nucleus, and in consequence it should be able to move very freely through matter. Its presence would probably be difficult to detect by spectroscope, and it may be impossible to contain it in a sealed vessel. On the other hand, it should enter readily the structure of atoms, and may either unite with the nucleus or be disintegrated by its intense field, resulting possibly in the escape of a charged hydrogen atom or an electron or both.[9]

It was an extraordinary feat, but actually the lecture received little attention outside the society, precisely because it was about prediction. Others simply could not see what Rutherford foresaw. Eleven years later, an American team led by Harold C. Urey of Columbia University found heavy hydrogen, calling it "deuterium." And light helium was discovered a few years later. At Cavendish, Chadwick devoted himself to looking for the neutral doublet. An extraordinarily patient man, he spent more than ten years at it, with some help from

Rutherford and some interference as well. "I just kept pegging away," he said later. "I did quite a number of silly experiments . . . I must say the silliest were done by Rutherford."[10]

Chadwick, who had lost some of his best years and his health during his Berlin internment, finally got lucky near the end of 1931, when he read of odd results in experiments in Germany and France—again, science advanced by openness and building on the work of others. This time, Chadwick read papers by two German scientists, Walther Bothe and Hans Becker, and the French husband–wife team of Irène and Frédéric Joliot-Curie, who published in *Comptes rendus de l'Académie des sciences*, the French equivalent of Britain's *Nature* and the United States' *Physical Review*.

Both teams had tried bombarding beryllium with alpha particles and produced radiation that they could not identify. Chadwick could. He repeated the German and French experiments and, on February 27, 1932, in *Nature*, published a groundbreaking paper: "Possible Existence of a Neutron." After ten years of variations of scattering experiments, Chadwick was able to shoot protons into paraffin, knocking out particles that were capable of punching through an inch of lead; that could happen only if they had the mass of a proton but no charge. It was that lack of charge that had made the discovery so difficult; protons (positively charged) and electrons (negatively charged) had been discovered and used as search vehicles to explore atoms and matter because their electric charges and trails could be followed. Chadwick's neutron also explained the differences between atomic numbers (1 to 92 at the time) and atomic weights (1 to 238). Uranium, for instance, had an atomic number of 92 because it contained 92 protons, but an atomic weight of 238 because it also contained 146 neutrons.

Why did Chadwick recognize the neutron when the others had not? Frédéric Joliot-Curie, who, with his wife, had produced some of the same results but was unable to understand them, answered, "The word neutron had been used by the genius Rutherford in 1923, at a conference to denote a hypothetical neutral particle which together with protons made up the nucleus. This hypothesis has escaped the attention of most physicists, including ourselves. But it was still present at Cavendish where Chadwick worked . . . Old laboratories with long tradition always have hidden riches. Ideas expressed in days gone by by our teachers, living or dead, taken up a score of times and then forgotten, consciously or unconsciously penetrate the thought of those who work in these old laboratories and, from time to time, they bear fruit: that is discovery."[11]

Chadwick himself affirmed that fact, talking of his many scattering experiments with Rutherford: "Before the experiments we had to accustom ourselves to the dark, to get our eyes adjusted, and we had a big box in the room in which we took refuge while Crowe, Rutherford's personal assistant and technician, prepared the apparatus . . . And we sat in this dark room, dark box, for perhaps half an hour or so, and naturally talked . . . It was those conversations that convinced me the neutron must exist."[12]

George Crowe, the technician, had his own stories of the way Rutherford worked in the laboratory, as he told Arthur Eve in describing the scattering experiments:

"Now, Crowe, put in a 50-centimeter screen."

"Yessir."

"Why don't you do what I tell you?—put in a 50-centimeter screen."

"I have, sir."

"Put in 20 more."

"Yessir."

"Why the devil don't you put in what I tell you, I said 20 more."

"I did, sir."

"There's some damn contamination. Put in two 50's."

"Yessir."

"Ah, it's all right. That's stopped 'em! Crowe, my boy, you're always wrong until I've proved you right. Now we'll find their exact range."[13]

Chadwick's discovery led to hundreds of others around the world—again, that is how physics worked in what Rutherford called its "heroic age"—including concepts essential to nuclear fission and the making of atomic bombs. At Cavendish, researchers explored a new range of transmutation, beginning with firing a neutron into a nitrogen atom (atomic number seven) and knocking out a proton. That left six protons and the nitrogen atom had become a carbon atom (atomic number six).

But there were some bad years for Rutherford between the great Bakerian Lecture of 1920 and Chadwick's 1932 discovery of the neutron, the proof that the lecturer, his boss, had somehow seen what no one else could even imagine. Not only was Rutherford tied down or tied up with the rebuilding and administering of Cavendish after its wartime decline in men and missions, the scientist had become, perhaps, too famous for his own good, traveling the world accepting awards and making speeches in places far and near, from South Africa to New Zealand again. He was away from Cambridge as long as six months at a time, promoting science to government leaders and the public, serving on august boards and commissions—and inspiring young men and women to look into the wonders and mysteries of all that was around them.

Rutherford had become an international statesman and salesman of science. One of his achievements was restocking the university libraries of Belgium, which had been destroyed in the war. Another was working out a scheme to reopen and refinance the Vienna Institute, officially still an "enemy institution," by persuading the British Royal Academy to pay the Austrians for the radium supplied to British laboratories before the war.

All that was made easier by the fact that Rutherford was elected president of both the British Association for the Advancement of Science, in 1923, and the Royal Society, in 1925. But in his speech on assuming the presidency of the association, Rutherford made one indefinite prediction that would later haunt him, saying that atomic energy might never be harnessed for useful purposes. "This possibility of obtaining new and cheap sources of energy for practical purposes was naturally an alluring prospect to the lay and scientific mind alike," he said in his inaugural address in September 1923. "It is quite true that if we were able to hasten radioactive processes in uranium and thorium, so that the whole cycle of their disintegration could be confined to a few days instead of being spread over millions of years, these elements would provide very convenient sources of energy on a sufficient scale to be of considerable practical importance. Unfortunately, although many experiments have been tried, there is no evidence that the rate of disintegration of these elements can be altered in the slightest degree by the most powerful laboratory agencies."[14]

In 1925, the year he was elected to a five-year term as president of the Royal Society, Rutherford agreed to twenty separate appearances around England—and turned down sixty more. He produced only one paper longer than a page that year. He wrote none in 1926, and none in 1928.

Worse, perhaps, Rutherford "saw" something new that was not there. Most of the experimental work he had done over five years in the 1920s was an attempt to prove that the nuclei of atoms were actually atomic structures themselves. Inside the nucleus, he suggested, was another nucleus orbited by something like electrons, the whole thing a minute version of the pinhead in the cathedral. He was wrong. The scientific world accepted Niels Bohr's supposition that the nucleus was "a mush."

He was still Rutherford, though. He still attracted the best and the brightest to Cavendish. With Chadwick as an effective associate, Rutherford did rebuild and expand the laboratories, and he worked with younger students and researchers, guiding and driving men who won Nobel Prizes. After Frederick Soddy won the chemistry prize in 1921 and Niels Bohr won the physics prize in 1922, other Rutherford protégés, most of whom did their prizewinning work at Cavendish in the 1920s, won the awards in later years: Francis William Aston, chemistry, 1922; Paul Dirac, physics, 1933; Chadwick, physics, 1935; Georg von Hevesy, chemistry, 1943; Otto Hahn, chemistry, 1944; Edward Appleton, physics, 1947; Patrick Blackett, physics, 1948; John Cockcroft and Ernest Walton, physics, 1951; and Pyotr Leonidovich Kapitsa, physics, 1978.

The last three men, along with Chadwick, had a profound

effect on Rutherford's quest to "know" the nucleus. Rutherford was well into his fifties when they came to Cavendish, but he was able to listen and learn and to change his mind. Beginning in 1924, when many thought Rutherford as scientist was a spent force, Kapitsa, Walton, and Cockcroft became important factors in their professor's move away from table-top experiments and into the new physics of big machines, big spaces, and big industries, particularly the electric power industry. Finally, in his annual president's address to the Royal Society on November 30, 1927, Rutherford conceded that the future would belong to "Big Science." After summing up the year's work and financial statements, Rutherford said, "In the short time at my disposal I would like to make a few remarks on the results of investigations carried out in recent years to produce intense magnetic fields and high voltages for scientific purposes."[15]

Rutherford talked about the development of high voltages in medical care and the power industry. Hospitals were already using transformers that produced steady power for X-ray machines using five hundred thousand volts and more, and industrial tests had produced five million volts and more. Power companies, he continued, had their own laboratories to test their own equipment, their transformers and insulators, under high voltages. Then he said, "The advance of science depends to a large extent on the development of new technical methods and their application . . . From the purely scientific point of view interest is mainly centered on the application of these high [industrial] potentials to vacuum tubes in order to obtain a copious supply of high-speed electrons and high-speed atoms . . . This would open up an extraordinarily interesting field of investigation which could not fail to give us information of great value, not only in the constitution of atomic nuclei but in many other directions."[16]

That was an admission of sorts suggesting the possibility that the use of radium and other radioactive sources to study the structure of the atom might have reached a natural limit—which meant Rutherford's past work might be just that: past. Like the Bakerian Lecture in 1920, the speech seemed at first to have a short half-life, but it was a hinge in the history of science. Writing eighty years later, in 2005, a Cambridge-trained physicist who had become one of the preeminent science writers of his time, Freeman Dyson, said,

> Toward the end of the 1920s, nuclear physics got stuck. Major mysteries remained to be solved . . . But it was hard to think of exciting new experiments that could be done with existing tools. The next round of experiments were minor variations of experiments that had already been done. Rutherford announced in London in 1927 that new tools were needed if nuclear physics were to move ahead . . . The most promising new tool would be a particle accelerator, an electrical machine that could produce a beam of artificially accelerated particles would be better than beams produced by radium. Artificially accelerated particles would be better than natural particles in three ways: they could be produced in greater quantities, they could have higher energies and they would allow experiments to be designed more flexibly . . . The switch from natural sources to accelerators would start a new era in the history of science.[17]

The pace of change in science over less than twenty years was its own accelerator. In 1911, Rutherford's determination and intuition made him the young champion of a new physics challenging the mastery of every thinker from the Greeks to Isaac Newton. But by the mid-1920s, many thought of him as the master of a classic physics defined by its small scale, focusing all study on the structure of the nucleus. In a clever *New York Times* review of Edward Andrade's 1924 book *The Structure of the Atom*, which was dedicated to Rutherford, the reviewer, Benjamin Harrow, began, "The trouble with Einstein and his relativity is that the earth is much to small for him; the trouble with Rutherford and his atom is that the earth is much too big for him. Relatively speaking, the atom is to the earth what the earth, perhaps, is to the cosmos. as a whole."[18]

There were many differences between the two great men, the experimenter looking into things unseen and the theoretician looking out at others unseen. But as Einstein worked at refining the formulas that he conceived at the beginning of the twentieth century, Rutherford at the age of fifty-six became a leader of the newest new nuclear physics. Welding and miles of wiring were about to replace sealing wax and string at Cavendish. Much of what Rutherford said about a new era that night in 1927 had come from work and conversations with Kapitsa, a charismatic twenty-seven-year-old Russian electrical engineer, son of a czarist general, who had come to Cavendish in 1921 as part of a Soviet mission to buy laboratory equipment—and just stayed on. 3

When the Russian first approached Rutherford about staying, the director said no. Kapitsa then asked how many researchers Cavendish had. "Thirty," said Rutherford. "What is the customary accuracy of your experiments?" Kapitsa asked. "About two or three percent," the director answered.

"Well, then," said Kapitsa, "one more student would not even be noticed within that accuracy." Rutherford laughed. He let him stay.[19] *

The young Russian was a Big Science man, sure that with the right backing he could develop new, powerful industrial-style tools to probe matter, specifically by creating intense magnetic fields with industrial dynamos. Rutherford provided the backing. He was quite taken with the young man and his talent. The first work assigned to new Cavendish researchers was a basic course, taught by Chadwick, in a tiny attic room called "The Nursery." Some men and women spent six months there. The young Russian completed the course in just two weeks. Within a year, Kapitsa had done some successful cloud chamber photography of alpha particles in intense magnetic fields. Within a couple of years, Rutherford created a special professorship in magnetic research for Kapitsa. But there was always one condition written into those agreements: Kapitsa, a great talker who traveled back and forth to the Soviet Union over the years, was not allowed to talk about politics. "No propaganda in my labs," blasted the boss.[20] Kapitsa, a loyal Russian but no communist, kept the promise, though he retained Soviet citizenship and knew some of the leaders of his country, including Leon Trotsky and Nicolay Ivanovich Bukharin, the editor of *Pravda*. Of the many pleasures of Kapitsa's dual life was a simple introduction he made in London: "Comrade Bukharin, I'd like you to meet Lord Rutherford."[21]

---

* This story, repeated over and over in Rutherford lore, may be apocryphal. James Chadwick, who was close to both men—Kapitsa was the best man at his wedding—always insisted it never happened.

Kapitsa was a charmer. "He had a touch of genius: in those days before life sobered him, he had also a touch of the inspired Russian clown," wrote Snow,[22] who knew him well. One year, on Rutherford's birthday, the young Russian gave the boss a gold mechanical pencil to replace the little stubs of wood and lead that Rutherford kept in his pockets along with notes on scraps of crumpled paper. Nice idea, but the new instrument was never seen again. 4

Kapitsa spent fifteen years with Rutherford. An accomplished electrical engineer, he did create the most powerful magnetic fields yet known—in fact, their power was not matched again until 1956—but nothing scientifically significant came of it in the study of matter. Kapitsa eventually shifted to cryogenics, low-temperature study of gases, creating the first liquid helium after he left Cambridge. His greatest achievement at Cavendish was persuading Rutherford to allocate large amounts of the budget to build his own laboratory, complete with the kind of big equipment never seen there. Big generators, transformers, and government grants—this was the new science, and Rutherford was now part of it, writing in an application for government funding, "I have great confidence in the ability of Dr. Kapitsa both on the theoretical and practical side, and he is the only man I know capable of carrying out such a difficult research . . . with good fortune, such an investigation cannot fail to yield results of great scientific interest and incidentally of possible industrial importance."[23]

The Russian, who had lost his wife and two children to influenza during the chaos of the Russian Revolution, survived on boundless and lively joy. His "Kapitsa Club," by invitation only, held weekly meetings in his rooms—a serious intellectual forum for young researchers to discuss their work in an informal setting. Kapitsa was popular with his colleagues, though some resented the way he charmed and flattered the boss to get his way. His own equipment, extraordinarily expensive for Cavendish, was installed in a laboratory built to his specifications in 1929 in the courtyard behind the main entrance to Cavendish. It was said, of course, as it had been before about Bohr and Moseley, and would again about other young men, that Rutherford treated Kapitsa as the son he never had.

Kapitsa called Rutherford "The Crocodile," at least in letters to his mother and to his second wife, saying that, like the fearsome river animal, "The Prof" always moved forward, never back. "In Russia the crocodile is the symbol for the father of the family," wrote Kapitsa. "It is also regarded with awe and admiration because it has a stiff neck and cannot turn back. It just goes straight forward with gaping jaws—like science, like Rutherford."[24] Kapitsa had a gift for knowing what Rutherford wanted to hear, including this: "Good work is never done with someone else's hands. The separation of theory from practice, from experimental work, and from practice above all, harms theory itself."[25] His letters, written between 1921 and 1935, are long and often perceptive about both science, big and small, and about The Crocodile, who was getting bigger all the time. One Rutherford anecdote was told by C. P. Snow, who was in a local tailor's changing room when Rutherford came into the shop, roaring, "This shirt is too tight around the neck. I'm growing in girth every day." Then he added: "In mentality, too!"[26]

One of Kapitsa's early letters to his mother reads, "Rutherford is satisfied, as his assistant told me . . . Whenever we meet, he greets me with kind words. On Sunday he invited me to tea at his house and I observed him at home. He is very homely and kind . . . But when he is displeased, then hold on. He will so go for you that you will not miss the point. But he has a marvellous mind! He is a quite specific intelligence; a colossal feeling and intuition I could never imagine such a thing previously . . . He is an absolutely exceptional physicist and a very original man."[27]

Kapitsa also wrote this:

Many admire Rutherford's extraordinary intuition, which, figuratively speaking told him each time how to set up the experiment and what to look for. Intuition is usually defined as an instinctive process of the mind, something inexplicable which subconsciously leads to the correct solution. In my view this may be partly true, but at any rate it is strongly exaggerated. The ordinary reader is simply unaware of the colossal work done by scientists. He only hears of those specific works which have produced results. Anyone who has closely observed Rutherford can testify to the enormous amount of work he did. Rutherford worked incessantly, always in search of something new. He reported or published only those works which had a positive result; these however constituted barely a few percent of the whole mass of work.

You had to talk to him only about fundamental facts and ideas without going into technical details . . . I remember when I had to bring him for approval, my drawings of the impulse generator for strong magnetic fields, for politeness sake, he would put them on the table, without noticing they were upside down and he would say to me: "These blueprints don't interest me. Please state simply the principle on which the machine works.[28] ↲

Kapitsa was shrewd enough to combine his flattery with teasing. He had become confident in the director's presence. Many were not. At high table at Trinity College one night, Kapitsa began talking about the book *Genius and Madness* by Cesare Lombroso. Rutherford listened quietly for a while.* Then, as the Lombroso discussion continued, he suddenly boomed out a question: "In your opinion, Kapitsa, am I mad, too?"

"Yes," said the Russian, who was anything but intimidated that night. "You remember a few days ago you mentioned to me that you had a letter from the USA from a big American company?" (It was apparently General Electric.) "In this letter ⤳ they offered to build you a colossal laboratory in America and pay you a fabulous salary. You only laughed at the offer and refused to take it seriously. I think you will agree with me, that, from the point of view of the ordinary man, you acted

---

* As he became older, surrounded by an essentially literary community, Rutherford showed himself to be a well-read man with interests well beyond laboratories and results. As had the librarians at McGill and Manchester, C. P. Snow learned some of that when the Cavendish director stopped him on a Cambridge street a few weeks after his first novel, *The* ↲ *Search*, was published in 1934 and said, "Congratulations, Mr. Snow. Let's take a walk." Then the man of science dissected the book, character by character, sentence by sentence—"I didn't like the erotic bits, I suppose it's because we belong to different generations"—and did it again with other books. (C. P. Snow, *A Variety of Men* [New York: Scribner, 1966], 12.)

like a madman."[29] Flattered, I suppose, Rutherford filled the hall with his booming laughter.

Later that year, Kapitsa's boldness, or his naïveté, caught up with him when he was denied permission to leave the Soviet Union after one of his Russian vacations. On October 10, 1934, Kapitsa's second wife, Anna, appeared at Rutherford's door. She said that the Soviets, despite past promises, had detained her husband. They believed he could combine his science and engineering skills to bring modern electric power technology to the Soviet Union. The Russians wanted to electrify their vast country, and they thought Kapitsa could do that.

Rutherford spent months pushing political buttons and pulling at political strings in several countries to make his argument that science was international. Kapitsa, he said, could not continue his work in the Soviet Union, particularly when he was obviously not in great mental shape when he realized he was essentially a hostage. But The Crocodile failed. He may have suspected, but he never knew, that Joseph Stalin personally had ordered that Kapitsa never be permitted to leave the Soviet Union. One of the letters that Rutherford received during this period was from the Soviet ambassador in London, Ivan Maisky, who said, "Cambridge would no doubt like to have all the world's greatest scientists in its laboratories, in much the same way the Soviet Union would like to have Lord Rutherford and others of your great physicists in her laboratories."[30]

The game was up, and Rutherford knew it. For months he had kept the matter secret, but on April 24, 1935, a London tabloid, the *Star*, headlined the story like this:

RUSSIA TO KEEP HIM
Halt to Cambridge Studies

Rutherford then began negotiations to have Kapitsa's laboratory and equipment shipped to Russia, as specified in Kapitsa's contract. This was Rutherford's little revenge. He was a canny fellow when he had to be; he had watched the sharp trading on the frontier when he was young. Cavendish did ship a duplicate of Kapitsa's Mond Laboratory—named for a British industrial family that had essentially financed the facility—brick by brick, including the front wall symbol of a crocodile, to Russia. The Soviet government paid thirty thousand pounds sterling for equipment and work that had cost less than half that to construct, dismantle, and then ship. And truth be told, science did not wait for politics. New advances in heavy electrical machinery had made the lab and most of its equipment obsolete by the time it reached Moscow.

There was such sadness in it all. In one desperate letter to Rutherford, Anna Kapitsa said of her husband, "If he did not commit suicide in Russia during the last year it was not for love of me or the children, it was only for love of you, not to let you down after all you did for him, after all the trust you put in him."[31]

Realizing that the cause was lost, and that Kapitsa was in a deepening depression, Rutherford wrote again and again to his friend, telling him to go back to work, that being in a laboratory was his only hope. Finally, Kapitsa did that, accepting a position as director of the Institute of Physical Problems of the Soviet Institute of Science. In 1978, he was awarded a Nobel Prize in Physics for his work in magnetism and in the production of liquid helium, based on experimentation he had begun at Cavendish decades before. He had already lost his position at the Soviet Institute, in 1946, when he refused to work on nuclear weapons development, but he was restored at the institute a year after Stalin's death. He was allowed to return to Cambridge for a visit in 1961. As he walked in for a dinner at the high table of Trinity College, he realized he had no academic regalia, a requirement there. The building's butler walked up to him and asked, "Are you Dr. Kapitsa?" and then handed him the robe that he had been holding since the Russian last wore it in 1934.

The man in charge of packing up the Mond Laboratory in 1935 was John Douglas Cockcroft, Kapitsa's assistant at Cavendish. Cockcroft, a Yorkshireman who had spent four years at war and then worked in the electrical industry, was already twenty-seven years old when he came to Cavendish in 1924. He had been studying mathematics at the University of Manchester before 1914 and had been inspired by Rutherford's lectures, though he had never met the great man. After the war he went to work as an engineer at Metropolitan-Vickers of Manchester, England's dominant electrical engineering firm. It was Metrovick, as it was called, that encouraged Cockcroft to go back to school and helped pay for it. He studied math again but was soon drawn into Rutherford's orbit, signing on as a PhD candidate at Cavendish.

Ernest Thomas Sinton Walton, an Irishman who felt he had learned all they could teach about science at Trinity College, Dublin—his master's thesis was on theories of hydrodynamics—arrived at Free School Lane asking to see

Rutherford, but the boss was busy, so Walton was sent upstairs to "The Nursery" to see Chadwick.

Rutherford, Cockcroft, and Walton worked together for five years. The Crocodile and "the boys," as Rutherford called the other two, entered the great race of Big Science, the international contest to split the atom on an industrial scale.

CHAPTER 7

I n 2004, a British journalist, Brian Cathcart, wrote a book
titled *The Fly in the Cathedral*, on what happened in the
five years after Rutherford, Cockcroft, and Walton came
together in 1927 on Free School Lane. "The last true gentlemen scientists," Cathcart called the three of them, "in a
grubby basement room at the famous Cavendish Laboratory
in Cambridge."[1]

It is a wonderful story, well told. Walton, with no real experience in nuclear physics, came to Rutherford, who had spent
his life chasing the fly, the nucleus, and said he was interested
in particle acceleration. It was late October 1927. Rutherford
was too busy to see Walton right away. Actually, he was
preparing his Royal Society speech calling for just such
research as a way of discovering secrets of the nucleus. The
younger man, of course, had no idea of that. The twenty-
four-year-old son of a Methodist minister in the Irish Free
State, Walton knew nothing about particle physics, except that
it was where the excitement was and Cavendish, at least in

Britain, was the center of that excitement. Walton's work at Trinity College, Dublin, had involved the flow of water around cylinders. That work won him an Exhibition of 1851 scholarship, just as Rutherford had won one more than thirty years earlier for radio wave experiments. When Walton met Chadwick, the deputy director liked what he saw, particularly Walton's photographs of his work.

"Follow me, my boy," said Rutherford when he finally met Walton,[2] taking him through the maze of courtyards and stairways that was Cavendish. They ended up in a large basement room with three workbenches. Cockcroft was assigned one but began spending most of his time with Kapitsa. The second was used by Thomas Allibone, another Yorkshireman who had come to Cavendish from Metrovick after talking with Rutherford about the possibilities of accelerating electrons with high voltages. "Bones," as Allibone was called, did not have much luck experimenting in his three years in the laboratory before he went back to Metrovick as the director of its laboratories. But he would not soon be forgotten at Cavendish for his ritual electrocution of a tame rat in the name of safe science. Rutherford had him do that—putting a hole clean through the animal—to demonstrate the dangers of working with high voltages. And high voltage was the conventional wisdom of the day. It was generally accepted that it would require five to eight million volts to accelerate particles to speeds that could penetrate the protective electric shields that were believed to coat and guard nuclei.

It was Cockcroft, reading a paper published by George Gamow—a Russian physicist who had studied under Rutherford before working in Germany and the United States—who realized it might be possible to penetrate the nucleus at much lower voltages. Gamow, a theoretician, considered the alpha particle a wave and suggested that, even with voltages of less than five hundred thousand, those particles were capable of "tunneling" their way out through the electric barrier around nuclei. Cockcroft's insight was this: if alpha rays could tunnel out, perhaps accelerated protons could tunnel into the nucleus. He calculated that using accelerated protons to penetrate the nucleus might require as little as three hundred thousand volts.

Once Cockcroft had persuaded Rutherford of this possibility, everything changed in Walton's basement lab. Cavendish had no room for the huge equipment needed to generate millions of volts, but it did seem possible to produce up to three hundred thousand volts in smaller spaces. So Walton and Cockcroft began to work together, with help from Metrovick, which agreed to manufacture test equipment that could fit through the historical entrance and winding corridors of Cavendish. By the end of 1928, the first large piece of equipment, a transformer, was wiggled into the rigid confines of their basement. Then came condensers and rectifiers to multiply voltages and change the power company's alternating current to the direct current necessary for laboratory work.

Since Rutherford's first scattering experiments and then his attack on nitrogen atoms, he had been known in the popular press as "the man who split the atom." That was great for headlines, but Rutherford knew that all he had managed to do so far was chip away at bits of nuclei. By the beginning of 1930, a grim international competition was on to build the first effective particle accelerator, which was really a race among the most powerful nations to split the atom, explode it for the first time. The idea was to bombard nuclei with particles accelerated by huge electrical machinery, break them open and . . . well, no one knew, actually. The idea was just to do it and see what would happen. The whole world was watching, some people a bit nervously. The *New York American* and other Hearst papers in the United States editorialized, "A colossal catastrophe might ensue. Will this planet, twirling peacefully a million years, be blown to smithereens?"[3]

There were five contenders, research laboratories, in the international race chronicled by Cathcart and others:

1. In their Cavendish basement, Cockcroft and Walton hoped to use protons (hydrogen nuclei) accelerated to higher and higher speeds by using three hundred thousand volts shot through tubes and pipes into collision with light metal foils, beginning with lithium foil, the soft metal that was the lightest of the solid elements.

2. At the Carnegie Institution in Washington DC, Merle Tuve, a small-town boy from Canton, South Dakota—by way of the University of Minnesota, Princeton, and Johns Hopkins—was using a transformer in oil, an apparatus so big that it had to be built outdoors. It stood, incongruously, on the Washington Mall near the Smithsonian Institution.

3. At the University of California, Ernest O. Lawrence, Tuve's best friend back in Canton, was trying to do it with his "cyclotron," a small circular accelerator.

4. In Pasadena, at the California Institute of Technology, Charles Lauritsen, was well equipped (particularly in comparison with the Cavendish team) with state-of-the-art equipment and 750,000 volts of alternating current supplied by the local power company, Southern Califor-

nia Edison. There and in other places, what was of primary interest to the power companies was the development of high voltages for hospitals' X-ray work.

5. Finally, there were the Germans Arno Brasch and Fritz Lange, backed by the Swiss engineering firm Brown Boveri. They had started out in 1927 trying to harness the power of lightning with cables strung between five-thousand-foot-high mountains in the Swiss Alps, until their partner, Kurt Urban, was killed in a fall and their work was moved back inside the laboratories of the Kaiser Wilhelm Institute in Berlin.

It was a great game, with great problems for all the players. Some of the British team's problems were solved, or at least made easier, when Rutherford was able to commandeer more space at Cambridge—Lecture Hall D in the Balfour Library—allowing his "boys" to bring in larger transformers.

That was late in 1930, a time of more honor and profound sadness for the Cavendish director. In December of that year, the Rutherfords' only child, Eileen, who was married to a young Cambridge mathematician, Ralph Howard Fowler, died giving birth to her fourth child. Then, two weeks later, the grieving professor was named a baron, choosing his own title—"Lord Rutherford of Nelson"—and designing his own coat of arms, a fanciful concoction connecting New Zealand and the universe: underneath a kiwi bird, there was a Maori warrior on the right and, on the left, Hermes Trismegistus, a mythical god who threw the planet Uranus into space. The motto read, *Primordia Quaerere Rerum*, "To seek things first." Rutherford cabled his mother back in New Zealand: "Now Lord Rutherford. More your honour than mine."[4]

Rutherford already had more titles than a man needed, and there were those who thought he was never the same again after Eileen's death. His wife, to others at least, was a cold and insecure woman, and their relationship with Eileen had become a bit strained, partly because the younger woman had a drink now and then. Mary Rutherford was particularly upset because her daughter and her friends, young marrieds, would picnic on the grass near the river Cam; anyone who passed by could see wine bottles in the picnic baskets.

Whatever the troubles were in the Rutherford home, they were not much different from most generational disputes. Eileen had different ideas—she and Fowler shared a house with another married couple, for example—that sometimes clashed with Lord Rutherford's sense of position and Lady Rutherford's strict propriety. Mary had come a long way from Christchurch, but she was nowhere near as comfortable as her gregarious husband and tried to mask her insecurities by having her way in all things small. She was noted both for hustling across rooms if she saw her husband touch a wineglass and regularly pointing out his famously sloppy eating habits to guests at their table: "Ern, you're dribbling again."[5]

Rutherford got old faster after that; at least, his body did. C. P. Snow, who knew him in those days, later wrote, "He was a big rather clumsy man, with a substantial bay window that started in the middle of his chest. I should guess that he was less muscular than at first sight he looked. He had large staring blue eyes and a damp and pendulous lower lip. He didn't look in the least like an intellectual . . . He worked hard, but with enormous gusto; he got pleasure not only from the high moments, but also from the hours of what to others would be drudgery, sitting in the dark counting the alpha scintillations."[6]

After Eileen's death, Rutherford spent a good deal of his time with his four grandchildren, and he became closer, too, to his son-in-law, who was part of a golfing foursome dominated by the big man's voice, if not his swing. Snow may have overestimated Rutherford's patience for counting scintillations, but the laboratory was always where he was happiest, his main escape—as it is more often than not for many scientists. The world makes more sense in there, and only a chosen few are allowed in, communicating in a language of their own, a mystery to most outsiders.

By then, Rutherford had slowed his manic traveling and was back to older (or younger) ways, both cheering on and badgering his men for results, particularly Cockcroft and Walton. The director had been at this for a long time and understood that many experimenters love to refine and talk about ideas, equipment, and process, but put off the big moment when they actually have to test those things. He made a joke after one long trip about the slow perfectionism of C. T. R. Wilson, whose result, the cloud chamber, did pass all tests and changed all subatomic research: "When I left I saw my old friend Wilson grinding a glass joint. I just stopped by and saw him still grinding the same joint."[7]

Rutherford suspected, or pretended he did, that Cockcroft and Walton were doing some procrastinating, putting off the day of reckoning, by testing and retesting, then spending most of 1930 remaking equipment to raise their voltage potential to eight hundred thousand from the three hundred thousand that they originally thought would do the job. The director's own sense of goal and urgency was not helped by reports that the Germans were working with voltages up to 2.4 million. American journals reported that Tuve at Carnegie was testing

at 1.9 million. Then came a *New York Times* mention of an
Ernest Lawrence speech on September 19, 1930, to the
National Academy of Sciences, in which he said his prototype
"cyclotron" was working: "Preliminary experiments indicate
that there are probably no serious difficulties in the way of
obtaining protons having high enough speeds to be useful for
studies of atomic nuclei."[8]

All the contenders in Cathcart's chronicle ran into more
and more trouble as they got closer to the goal. Big Science
was simply too big for old laboratories. Cockcroft and Walton's fifteen-foot-high tower was too tall for a basement
room; Tuve's work had to be done outdoors, and his calcula-
tions were thrown off by dust and insects. Lawrence was using
a discarded eighty-five-ton electromagnet originally made for
a telephone company, but Berkeley had no space big enough
to hold it. The Americans were issuing optimistic statements,
but the Cavendish team was nowhere near ready to pro-
nounce progress. Late in 1930, in a letter to his fiancée,
Winifred Wilson, Walton wrote, "The rate of progress in the
lab has been zero lately. I spent the whole of last week looking
for a very small leak in a complicated piece of apparatus. In
the end I had to take it all to pieces. It is now assembled again
and I hope that it will work more satisfactorily."[9]

In the end it took Cockcroft and Walton almost two years to
fill Lecture Hall D with a collection of devices that looked like
the set for a 1930s Buck Rogers movie. As well it should have,
because it was photos of labs like theirs that inspired the writers and filmers of science fiction. The *Illustrated London News*
photographed the accelerator machinery. The caption listed
the devices: "A is the transformer, B is a condenser, C the four-

stage glass rectifier tower, D the acceleration tube, E the observation hut (with black curtain drawn) and F the spark gap spheres. The two unidentified towers in the background are another condenser set (next to the spheres) and the transformer for the proton supplier apparatus (far right)."[10]

Three things could not be shown in the picture. One was the control table with throw switches, gauges, and bulbs attached to wires running everywhere. Another was a big metal "top hat," looking as if it were left over from a New Year's party for giants. And, Cavendish-style, it all cost just over one hundred pounds. Great stuff, and very dangerous. Cockcroft and Walton often had to crawl around the lab, because anyone standing up when the switches were thrown could end up like Allibone's rat.

The top hat hid the proton supply, a tank of hydrogen gas pumped into a vertical glass tube, three feet long, with an anode at the top and a cathode at the bottom. In the simplest terms, a forty-thousand-volt jolt from anode to cathode broke up the hydrogen atoms, sending huge numbers of the hydrogen nuclei (protons) shooting down through shiny pipes inside the fifteen-foot-high accelerator tower, all in a vacuum, and powered by four hundred thousand volts of direct current. The protons in that dense beam of protons, fifty thousand trillion per second—electrons were thrown aside by the high voltage—sped up to thousands of miles per second before they hit the lithium foil target at the bottom. (Not all those protons hit or could be counted; the experiment was another exercise in statistical probability.) The foil was at a forty-five-degree angle and thousands of nucleus-to-nucleus collisions turned the particles toward a phosphorescent screen. The man in the little darkened and lead-lined wooden hut was usually Walton, looking into a microscope, ready to calculate results. The "hut," in true Cavendish style, was actually a tea chest, one of the crates used to ship tea leaves from Asia to London. The same kinds of chests ended up floating in the harbor after the Boston Tea Party before the American Revolution.

The year 1932 began triumphantly for Cavendish, with the February announcement that Chadwick had found the ever-elusive neutron. *Time Magazine* in the United States reported the accomplishment with both respect and the detached amusement of folks wondering what it is the British were doing over there, saying in its own choppy argot, "There most probably is a neutron, smallest bit, last resolvable particle of Matter . . . Dr. James Chadwick of Cambridge University's Cavendish Laboratory, brightest spot of British science, declared for the existence of neutrons. Ernest Baron Rutherford, director of the Cavendish, confirmed the investigation. And no brash statements ever come from Professor Rutherford, 1908 Nobel Laureate, the man who established the existence and nature of radio active transformations, the electrical structure of all matter, the nuclear structure of atoms."[11]

However, reports continued to come across the Atlantic that one way or another the race to split the atom would go to the newest, the biggest, the swiftest of Big Science, the Americans. In *Physical Review*, Robert van de Graaff, a Princeton physicist who had been inspired by Rutherford's 1927 speech and was working with Tuve, wrote of his 1.5-million-volt generator, saying, "The application of extremely high potentials to discharge tubes affords a powerful means for the investigation of the atomic nucleus and other fundamental problems."[12] Tuve, in Washington, wrote in the same journal, "We hope in the near future to undertake a program of quantitative measurements using the newly developed Van de Graaff generators."[13] The ever-optimistic Ernest Lawrence, with a new partner, Stanley Livingston, declared that his cyclotron had produced 1.1 million volts and that his operation "has now been brought to a stage of development where it can serve in experimental studies of atomic nuclei."[14]

Finally, in a February 1932 issue of *Nature*, Cockcroft and Walton did come up with a four-paragraph letter on their equipment, including the expectation that they could produce eight hundred thousand volts.[15] Lawrence and Livingston were in *Physical Review* a month later, saying that their cyclotron had reached 1.22 million volts and a newer version in the works would reach 10 million volts. Their work, they said, was "probably the key to a new world of phenomena, the world of the nucleus."[16] But neither side had split any atoms.

CHAPTER 8

G et on with it!" Rutherford roared,[1] charging into Cockcroft and Watson's crowded landscape of towers and transformers in Lecture Hall D one day early in April 1932. Chadwick had been there a few minutes before and saw that the experimenters had a steady beam of protons flowing through their glass acceleration tower, but they were still measuring and testing. Accounts vary as to how angry the director actually was—"You're wasting your time," he said— but he certainly got the message across. 3

On April 14, 1932, Walton began the first run, while Cockcroft was working with Kapitsa in his lab. Alone in the clanking, buzzing, hissing, and zapping of the lab, Walton recalled later, "When the voltage and the current of protons reached a reasonably high value, I decided to have a look for scintillations. So I left the control table where the apparatus was running and I crawled over to the hut under the accelerating tube. Immediately I saw scintillations on the screen. I then went back to the control table and switched off the power to

the proton source. On returning to the hut no scintillations
could be seen. After a few more repetitions of this kind of
thing . . . I then phoned Cockcroft who came immediately.*
He had a look at the scintillations and after repeating my
observations he was also convinced of their genuine character. He then rang up Rutherford who arrived shortly afterwards."[2]

The next problem was only slightly scientific. No one
wanted to watch the boss, with his bulk and his bum knee,
crawl across the lab into the little darkened hut. So they
turned off the machinery, and Rutherford walked to the cur-
tained tea crate. "With some difficulty we manouvered him
into the rather small hut and he had a look at the scintillations. He shouted out instructions such as 'switch off the proton current'; 'increase the accelerator voltage', etc. but he said
little about what he saw. He ultimately came out of the hut,
sat down on a stool and said something like this: 'Those scintillations look mighty like alpha-particle ones. I should know
an alpha particle scintillation when I see one for I was in at
the birth.'"[3]

The atom had truly been split by man for the first time.
What they saw that day was lithium nuclei of mass seven hit
by a proton of mass one and disintegrating into two alpha
particles (helium nuclei) of mass four. The lithium atom was
split. Hydrogen and lithium had been turned into helium.
There was one other important result: the total mass of the
end product was different from the original mass of the
hydrogen. Some of the mass—0.02 unit of atomic weight—

---

* The telephones were new. Rutherford had resisted their installation for
years.

had been transformed into energy in the violence of the action. That was the number predicted by the formula $E = mc^2$. The energy produced was equal to the mass multiplied by the square of the speed of light. It was the first experimental proof of Albert Einstein's 1905 theory of relativity. $3 *$

"For the first time," wrote Cathcart in his narrative, "a man-made apparatus had penetrated the nucleus and better than that, shattered it."[4] Freeman Dyson, reviewing Cathcart's book, wrote of the great competition of the Americans and the big man from New Zealand: "The Americans had the better machines, but Rutherford had a more single-minded concentration on the scientific goal . . . Van de Graaf [sic] and Lawrence were the hares. Rutherford was the tortoise. The tortoise won the race."[5]

Rutherford's first reaction was to swear Cockcroft, Walton, and Chadwick to secrecy about what had happened until the results could be published in *Nature*. Only God could know what the Americans would come up with if they knew in advance of publication. In the two weeks before publication, Cockcroft and Walton used the time to repeat the experiment with fifteen other elements. Boron, fluorine, and aluminum disintegrated in the same way as lithium. All of the new samples—from the lightest, beryllium, to the heaviest, uranium—produced some alpha particles when they were bombarded by the protons.

Of course the secret did not really hold. Rutherford wrote an excited letter to Niels Bohr. Walton wrote to his fiancée about a "red letter day." On the Cambridge campus, a young physicist, Neville Mott, passed a note to a friend: "Li + H = 2He."[6] Lithium plus hydrogen became two helium nuclei. And then Rutherford, just before the *Nature* publication, changed the title of an April 28 speech to the Royal Society in London to "The Structure of Atomic Nuclei." At the same meeting where Chadwick spoke of the discovery of the neutron, Rutherford told members what had happened in Lecture Hall D. Then he swept his arm toward Cockcroft and Walton and boomed out, "Stand up, boys! Let everyone have a look at you!"[7]

The *Nature* article appeared the next day under the title "Disintegration of Lithium by Swift Protons":

> The brightness of the scintillations and the density of the tracks observed in the expansion chamber suggest that the particles are normal alpha particles. If this point of view turns out to be correct, it seems not unlikely that the lithium isotope of mass 7 occasionally captures a proton and the resulting mass of 8 breaks into two alpha particles, each with mass 4 and each with an energy of eight million electron volts. The evolution of energy on this view is about sixteen million electron volts per disintegration, agreeing approximately with that to be expected from the decrease of atomic mass involved in such a disintegration.[8]

If Rutherford had suffered through some bad times, he was certainly back on top. As C. P. Snow wrote after the Royal Society meeting and *Nature* publication: "The year 1932 was the most spectacular year in the history of science. Living in Cambridge one could not help picking up the human, as well as the intellectual excitement in the air . . . I mean an intellectual climate different in kind from anything else in England at the time. The tone of science was the tone of Rutherford: magniloquently boastful—boastful because the major discoveries were being made—creatively confident, generous, argumentative, lavish and full of hope. The tone differed from the tone of literary England as much as Rutherford's personality differed from that of T. S. Eliot."[9]

For those who had never heard of *Nature* or the other technical journals that were the network of world science, the Cavendish triumph reached the popular press on May 1, 1932. The story was broken by *Reynold's Illustrated News*. A reporter had heard about Rutherford's Royal Society speech and tracked him down outside a meeting in London. "Your conclusion is fairly correct," answered Rutherford rather unhappily, knowing that Cavendish was about to lose control of its own story. And so it did, as the Sunday edition of the *Illustrated News* led with a five-deck headline:

SCIENCE'S GREATEST DISCOVERY

THE ATOM SPLIT AT 100,000 VOLTS

Secret of Cambridge Laboratory

MAKING A NEW WORLD

ENERGY WITHOUT LIMIT[10]

"On the authority of Lord Rutherford, the world famous scientist, Reynold's is able to announce exclusively that years of patient experiment at the Cavendish Library at Cambridge have at last been successful," read part of the lead paragraph. The next morning the laboratory on Free School Lane was surrounded by reporters, photographers, and correspondents of the British Broadcasting Corporation and other representatives of the new world medium of radio. Cathcart wonderfully described the photograph flashed round the world of Rutherford and his two painfully shy "boys":

The result is a priceless image: Rutherford almost dapper in grey Homburg hat and dark suit with wing collar and watch-chain, Cockcroft with heavy bags under his eyes but sporting a racy striped necktie and respectable three-piece, and Walton with the pockets of his shapeless tweed jacket where the single button is aching at the strain. All three wear expressions that combine happiness with embarrassment in roughly equal measure.[11]

The photograph accompanied one of the three pieces that *The New York Times* published in a week. The first was the paper's lead story on May 2, 1933, under the headline

ATOMS TORN APART,
YIELDING 60% MORE
ENERGY THAN USED

The story began, "LONDON, May 1—The atom has not only been split but an element has been transmuted into another element—atoms of hydrogen have been turned into atoms of that rare and commercially valuable gas, helium. At the same time the energy of part of an atom has been released in a quantity 60-percent greater than the amount used to produce this phenomenon." Then Rutherford was quoted: "Thus far the experiments have not yielded anything which would be of immediate commercial value . . . We cannot claim that for our experiments thus far, for the simple reason that for every particle of additional energy obtained it requires millions of particles to make it effective."[12]

Sir Leonard Hill, director of the London Light and Electrical Clinic, ignored the particle ratio and was quoted: "This discovery is the beginning of something far bigger and more important than the layman might imagine. For the first time in history we have got more energy out of something than we put in it. Every school boy learns the law of the conservation of energy—briefly that we cannot 'make' energy. To all intents and purposes this law may be considered broken."

The lead *New York Times* editorial and a Sunday analysis on May 8 was headlined "The Atom Is Giving Up Its Mighty Secrets." Ernest Lawrence, in New Haven for his wedding, must have seen that one. At any rate, he sent a telegram back to Berkeley that day: "Get lithium from chemistry department and start preparations to repeat with cyclotron. Will be back shortly."[13] Merle Tuve of Carnegie was there for the wedding. Fifty years later he remembered it all and said, "They caught us with our pants down."[14]

In the midst of the excitement, Walton wrote to his fiancée, Winifred Wilson, "I have learned one thing from reading the papers this week and that is not to believe all you see in them."[15] She wrote back that she was making an album of clippings and that their old school, Methodist College in Belfast, was giving pupils a half-day holiday to celebrate Walton's achievement. Three days after the announcement, Einstein happened to be visiting Cavendish and asked to meet Cockcroft and Walton and inspect their equipment. "He seems a very nice sort of man," Walton wrote to Winifred.[16]

Rutherford often seemed grouchier in his later years; certainly he was more impatient with others than ever, but he was still productive at the highest level of physics. In the years after the splitting of the atom, he produced seven significant papers on "heavy hydrogen"—the existence of which he had first suggested in his Bakerian Lecture of 1920—hydrogen isotopes with a mass number greater than one. They were shown to exist in 1933 by Harold Urey at Berkeley, and then produced there by G. N. Lewis as deuterium and tritium, part of what came to be called "heavy water." It was plain old H_2O, or at least 99.985 percent of it was; just 0.015 percent of the water contained the hydrogen isotope deuterium. That scientific curiosity would one day become critical as a "moderator" that could control the speed of slow neutrons used to bombard isotopes of uranium, making possible nuclear fission and the development of certain nuclear reactors and the atomic bomb. 3

Ironically, Rutherford was doing his best work now on paper, writing intuitive and suggestive theory. He was past sixty, and his hands shook enough that his own men, behind his huge back, called him a "thundering nuisance" in the laboratory, dropping equipment, breaking mica screens, and setting work back by hours and days of cleanup and reconstruction. Rutherford routinely destroyed photographic results by pulling them out of fixer pans too soon, dropping ashes on them, and getting the strips tangled in his feet.

However, there was nothing wrong with his mind, or his inspirational power—or with his native sense of what was true and what was not. He told the science writer Ritchie Calder that leaps of imagination were generally the province of young men, but older and wiser men had to be there with dry tinder to catch the new sparks thrown off by new generations. And Rutherford was famously there to do just that. He was in constant touch, in long letters, with the new generation, beginning with Lawrence at Berkeley; with the other American stars, Urey and Lewis; with Gamow; and, of course, with Bohr. He was also in touch with the Germans, beginning with Otto Hahn. Lord Rutherford, along with Einstein, the most revered of scientists, was still desperately trying to set science apart from the political madness associated with the rise of Hitler and Stalin.

In nuclear physics, all roads still led to Cavendish and Rutherford. It was often said over the years that he had more friends and fewer enemies than any man of his generation. There was some truth in that, but there was also reason and purpose. The younger Rutherford—the tireless traveler, recommender, and letter writer—was the personal force in the creation of an international scientific community, believing that private and public communications across national and academic lines were crucial to driving research forward. He understood that the greatest importance of new discoveries was that they opened new areas of experimentation. He worked at friendship and generous collaboration. He kept in touch as well as he could with former students during and after World War I, including Hahn and Geiger in the enemy's capital, Berlin. In those years, Geiger named his third son Ernst after "The Prof." 3

As that war ended, Geiger wrote to Rutherford, "I need hardly say that all that has happened these last four years has had no influence on my personal feelings for you, and I hope, dear Professor Rutherford, that you will still take a little interest in your old pupil who keeps his years in Manchester always in pleasant memory."[17] When Bohr won the Nobel Prize in 1922, he wrote to Rutherford, "I have felt so strongly how much I owe to you not only for your direct influence on my work and your inspiration, but also for your friendship in these twelve years since I had the great fortune to meet you for the first time in Manchester."[18]

Science, as Rutherford had said, was in its own heroic age. Creative men in other fields, who barely understood science, were also beginning to realize that something new was rising in these laboratories. "The experts," wrote Harold Nicolson in a 1933 novel, *Public Faces*, "had begun to whisper the words . . . 'atomic bomb' . . . that would destroy all matter within a considerable range and send out waves that would exterminate all life over an indefinite area."[19]

The world was turning sour. Financial depression was beginning to strangle developed economies, and one of the consequences was the rise of tendencies toward totalitarianism—violent totalitarianism in continental Europe, in the Soviet Union, and in Japan. In Germany, the great rival of Great Britain and the United States in physics and other sciences, the republic that had been created after World War I was collapsing. The National Socialist Party, led by Adolf Hitler, won 37 percent of the vote for the Reichstag in July 1932; the Nazis had become the largest and most powerful party in the country. By the end of the year, the country was in political chaos, and on January 30, 1933, Hitler became the chancellor of Germany.

Two months later, the Nazis passed the "Law for the Restoration of the Professional Civil Service," mandating that only "Aryans" could hold government positions, including faculty positions at German universities. "Non-Aryans" were defined as anyone with a single Jewish grandparent. And many of Germany's greatest scientists fell into that category, including James Franck, winner of the 1925 Nobel Prize in Physics; Max Born; Lise Meitner; Fritz Haber, who had believed it was his patriotic duty to help his country develop poison gas; and the former director of the Kaiser Wilhelm

Institute of Physics at the University of Berlin, Albert Einstein. In fact, Einstein, a German citizen, was a principal target of the racial laws. Two other German Nobel laureates, Johannes Stark and Philipp Lenard, were leaders of a "German physics" movement, which was busily engaged in denouncing what they called the "Jewish physics" of Einstein.

Stark and Lenard, obvious favorites of the new regime, also attacked "Jewish Aryans" or "white Jews," particularly Werner Heisenberg, the 1933 Nobel laureate in Physics, because they had decided that he "thinks like a Jew." Heisenberg survived the purge, partly because his mother and the mother of Heinrich Himmler, a principal Hitler deputy, had been friends since childhood. Whatever the reason, Himmler informed the gestapo to end its around-the-clock surveillance of Heisenberg, stating that he was too valuable to be liquidated.

By May 1933, fifteen hundred "non-Aryan" scientists had been dismissed from German universities and laboratories. At least a quarter of the country's physicists were forced out of their positions. William Beveridge, director of the London School of Economics, had just returned from a conference in Vienna, and after hearing of the dismissal of Jewish academics, he decided to try to create a network to help resettle and, if possible, find jobs for German professors and scholars. He made up a name: Academic Assistance Council. On May 6, 1933, Beveridge approached his country's most famous scientist, Lord Rutherford. About that mission, Beveridge gave this report:

> Most important of all I persuaded Rutherford after a first refusal on the grounds that he was up to his eyes in other work, and against strong opposition by Lady Rutherford, to

become President of the Council . . . In the end it was our cause not friendship that brought him over. As we talked he exploded with wrath at Hitler's treatment of scientific colleagues whom he knew intimately and valued. He would be miserable not to be with us if we went ahead. He did everything and more to make our going ahead possible.[20]

Some of Rutherford's friends were surprised that he took on the job and threw so much time and energy into it. He was an accomplished politician of science, finding backing and financing for promising men and projects. Women too. Though they competed and sometimes totally disagreed with each other's scientific conclusions, Rutherford and Marie Curie became friends and allies in one struggle after another, including the controversy over whether the measuring unit of radioactivity would be called the "curie." Rutherford stood with her, perhaps because they both began as outsiders—the little girl from Poland and the country boy from New Zealand—when the scientific establishment of the day, all male, refused her credit and honors for her own work, saying the real thinking must have been done by her husband or male assistants. And years later, he defended her again when she became a center of tabloid scandal. After Pierre Curie's death left her a widow, it was discovered that she was having an affair with a married colleague, Rutherford's Cambridge friend, Paul Langevin.

As skilled as he could be behind the closed doors of government, Lord Rutherford had little use for real politicians, and no interest at all in religion beyond knowing the words, if not the tune, of "Onward, Christian Soldiers." No one ever really knew his politics, if he had any. One of the times he talked about such things was a brief conversation with one of his young men, Marcus Oliphant, a socialist, after they watched a military tattoo, all flags and pageantry, on Salisbury Plain: "That expressed the true spirit of England far better than any nonsense you read in the New Statesman."[21]

In fact, however, Rutherford was ahead of Beveridge. He had already found positions and housing for some men he knew, taking on Max Born at Cavendish, then successfully nominating the German physicist for a chair at the University of Edinburgh.* There was, however, one German scientist whom Rutherford would not personally help or even meet: Fritz Haber. Said Born in a letter to Chadwick: "He violently declined an invitation to my house because Haber was to be present . . . He did not wish to have any contact with the man who had invented chemical warfare with the help of poison-gas."[22] 3/4 ✳

The council was formed before the end of the month, as Rutherford took on a task he usually avoided, raising money. He also took some heat from former German colleagues who supported the new regime. Johannes Stark, who had worked with Rutherford and Bohr years before, wrote to say, "In the name of the majority of German scientists, I appeal to you, as the leading representative of English science, to help in putting an end to the mischievous propaganda against Germany in English scientific circles, and in avoiding an estrangement between English and German scientists."[23] Rutherford answered, "This country has always viewed with jealousy any interference with its intellectual freedom,

---

* Born went on to win the 1954 Nobel Prize in Physics for his work on quantum physics.

whether with regard to science or learning in general. It believes that science should be international in its outlook and should have no regard to political opinion, creed or race ... We all sincerely hope that this break with the traditions of intellectual freedom in your country is only a passing phase."[24]

On October 3, 1933, four months after his first conversations with Beveridge, Rutherford chaired a mass meeting of ten thousand people in Royal Albert Hall to raise money for what he called "The Wandering Scholars." He began, "This meeting has been called to consider a problem of great magnitude and of ever increasing urgency—the problem of relief of German refugees ... Each of us may have his own private political views, but in this work of relief all such political differences of opinion must give way before the vital necessity of effectively conserving this great body of learning and skilled experience which otherwise will be lost to the world."[25] Then, with a huge swing of his arm, he boomed, "Ladies and gentlemen, my old friend and colleague, Professor Einstein."

Einstein gave the main speech, saying, "It cannot be my task to act as the judge of the conduct of a nation which has for many years considered me one of its own; perhaps it is an idle task to judge in times when action counts ... How can we save mankind? How can we save Europe from a new disaster? It is only men who are free to create the inventions and intellectual works which, to us moderns, make life worthwhile ... Without such freedom there would have been no Shakespeare, no Goethe, no Newton, no Faraday, and no Lister. There would be no comfortable houses for the mass of people, no railway, no wireless, no protection against epidemics, no cheap books, no culture, and no enjoyment of art for all."[26]

By the end of 1936 the council had aided thirteen hundred

of the wandering scholars, most of them Germans, most of them Jews. Many of them had lost everything, beginning with their academic positions. Their offices and their homes had been ransacked or destroyed by gangs of Nazi thugs—if they were lucky enough to have fled their countries with their families before things became worse. Einstein was in California when Hitler came to power. In March 1933, knowing he could never return to work in the new Germany, the Third Reich, Einstein resigned publicly from the Prussian Academy of Sciences. He did not know it, but the Prussian minister of education, Bernhard Rust, had planned to commemorate April 1, a day set for the boycott of all Jewish businesses, by expelling Einstein from the academy at a ceremony. Instead, Rust sent troops of young Nazis to ransack Einstein's home in Berlin.

In a time of economic depression in both England and the United States, and a time when both countries were also plagued by significant anti-Semitism, when many of the best universities in the world had restrictive Jewish quotas for both faculty and students, it was no small accomplishment for the ad hoc Academic Assistance Council to help make new lives for so many. After one of Rutherford's articles in the *Times* of London about the plight of Jewish scientists, his old friend from Manchester, the man who would one day become the first president of Israel, Chaim Weizmann, wrote to say, "You have my admiration and profound gratitude . . . Jews outside Germany have avoided expressing their feelings of indignation in case it upsets those whom, for the sake of peace, are trying to negotiate with the Germans . . . Your voice raised yesterday with so much dignity and restraint brought, I am sure, comfort and solace to many sufferers, and you have placed us under a great debt of gratitude."[27]

CHAPTER 9

Rutherford turned sixty-two in 1933. He was traveling abroad less, though he kept up his constant correspondence with Niels Bohr, with Otto Hahn, and with all the other "boys" he had helped along over the years in Manchester and Cambridge and Montreal. At home in England he was doing more speeches than experiments. He was still visiting Cavendish students and researchers, pulling up a stool and asking them the old questions about what they were doing. The questions were still simple and good: "What precisely are you doing? How? Why? Now let's see what the results are." And he had lost none of his powers of inspiration. Even more than that, according to Samuel Devons, who was at the laboratories in the 1930s, and in the 1950s was appointed to Rutherford's old professorship at Manchester,

The Cavendish was Rutherford's domain . . . His influence there seemed a wholly natural phenomenon. Benevolent guidance, leadership and intellectual authority flowed from him, and loyalty was returned. One would no more question his influence on those around him than one would that of the sun on the satellite planets. Rutherford, the Cavendish Professor, was the centre of light and warmth and life. It was the natural order of things. Young undergraduates were way out on the periphery of this constellation, but we could bask in the sunlight just the same.[1]

The lab visits were fewer, however, particularly compared with the weekly sessions of years past. "One would receive occasionally, perhaps once or twice a year, a more-or-less unannounced visit from Rutherford," said Devons.[2] Still, Rutherford got just as excited and involved as he had in the old days, particularly when he worked with his best people: Chadwick, of course; Cockcroft and Walton; and Mark Oliphant, working on hydrogen isotopes, who told this story:

Rutherford produced hypothesis after hypothesis, going back to the records again and again doing abortive arithmetic throughout the afternoon. Finally we gave up and went home to think about it. I went over all the afternoon work again, telephoned Cockcroft, who had no new possible ideas to offer, so I gave up and went to bed tired out . . . At three o'clock the telephone rang. My wife came back to tell me that "The Professor" wanted to speak to me. Still drugged with sleep I heard an apologetic voice express sorrow for waking me, then excitedly say, "I've got it! Those short-range particles are helium of mass 3." Shocked into attention, I asked on what possible grounds could he conclude that this was so . . . Rutherford roared, "Reasons! Reasons! I feel it in my water!"[3]

Oliphant raced to the lab and began running tests. The Professor's water was right.

More often than not, though, instead of prowling the corridors demanding results, Rutherford was taking longer vacations with his grandchildren and spending a good deal of his time preparing speeches for the most prestigious venues in Great Britain: a half dozen a year to the Royal Society, Watts lectures, the Boyle Lecture at Oxford, an annual address to the Chemical Society, the Mendeleev Centenary address, the Henry Sidgwick Memorial Lecture, and the Norman Lockyer Lecture.

In addition, Rutherford's "boys" were leaving Cavendish. In 1935, Blackett went to the University of London; Walton accepted a chair at Trinity College, Dublin; and Chadwick took a chair at the University of Liverpool, saying to friends, "It was becoming very difficult to push on without some new equipment. I couldn't go any further with what I had . . . It meant complicated equipment and Rutherford had a horror of complicated equipment."[4]

Rutherford recommended them all, but it meant his world was changing. Oliphant was offered a chair and his own laboratory at the University of Birmingham. Although Rutherford had recommended his old student and young friend from the Antipodes for the position, this time he showed the hurt when Oliphant accepted. "I'm surrounded by unfaithful colleagues," he roared in an angry scene, telling Oliphant, "Go and be damned."[5] Within an hour, Rutherford calmed down and tracked down Oliphant to apologize, promising him any help or equipment he needed at Birmingham—a promise the old man kept.

Rutherford was becoming something of a victim of his

own success, but if any man was a force of nature, he still was. As his biographer, David Wilson, wrote after Rutherford's boys heroically won the race to split the atom, "Superimposed on this background of comfortable, essentially kindly, totally non-commercial, English middle-class academic life, are a series of rapid, exciting discoveries. Experimental physics, led by Cavendish, produced in the decade of the 1930s the most astonishing ten years of development that science has ever seen. These discoveries brought nuclear physics from the most recondite 'pure' laboratory sciences into the atomic era, the age of nuclear power engineering and nuclear weapons."[6]

Writing about Rutherford's 1919 experimental splitting of the nitrogen atom, C. P. Snow wrote in 1966, "The rest of that story leads to the technical and military history of our time . . . He was a great man, a very great man . . . It was an astonishing career, creatively active until the month he died. His insight was direct, his intuition, with one curious exception, infallible."[7]

The exception that Snow wrote of was Rutherford's assertion that the power of nuclear energy might never be harnessed for good or evil. In September 1933, apparently responding to a recent article by Karl Taylor Compton, the president of the Massachusetts Institute of Technology, predicting that younger scientists were on the way to harnessing atomic energy, Rutherford repeated to the British Association a variation of what he had told the group in 1923—this time in more colorful language: "These transformations of the atom are of extraordinary interest to scientists, but we cannot control atomic energy to an extent which would be of any value commercially, and I believe we are not likely ever to be able to do so," he said. "A lot of nonsense has been talked about transmutations. Our interest in the matter is purely scientific . . . The energy produced by the atom is a very poor kind of thing. Anyone who expects a source of power from the transformation of these atoms is talking moonshine."[8]

Perhaps Rutherford only wished that were so, that the forces he discovered were best left in the hands of scientists. Thirty years before, after releasing the power of the forces binding atoms, he had talked about fools blowing up the world. In a 1916 London lecture, "Radiation from Radium," he had estimated that the intrinsic energy in a pound of radium would equal the energy of at least one hundred million pounds of coal, then added that he hoped that, before anyone discovered how to release such energy, men would have learned how to live in peace with their neighbors. He did not say "atomic bomb," but that's what he was talking about. And he had once said to Bohr, "Well, Niels, if in a nuclear reaction mass disappears, energy will appear, and ultimately, whatever its initial form, be degraded to heat. It might be used."[9]

In fact, it was revealed after World War II, long after Rutherford's death, that in the early 1930s he had specifically discussed atomic weapons with Sir Maurice Hankey, the secretary of the Committee of Imperial Defence. As an old man, Hankey spoke of being pulled aside by Rutherford at a committee meeting and being told, "The experiments on nuclear transformation which he was supervising at Cambridge . . . might one day turn out to be of great importance to the defence of the country. He did not quite know in what way this would be so . . . but some inner sense, for which he apparently saw no scientific justification, correctly warned him that someone should, as he put it, 'keep an eye on the matter.'"[10]

Then, in November 1936, Rutherford gave both the Henry Sidgwick Memorial Lecture and the Norman Lockyer Lecture. In the first, he talked of the work of the next generation of nuclear physicists, citing Enrico Fermi in Italy and Norman Feather at Cavendish, showing that "slow neutrons" were far more efficient than faster projectiles in entering atomic nuclei. Their power was in their mass: a thimbleful of neutrons might weigh as much as a hundred million tons. Though he did not say so directly, Rutherford obviously understood the potential of slow neutrons. The efficiency of particles without electric charge, which meant they could not be diverted by electromagnetism, could change the reality that, although splitting the atom released tremendous amounts of energy, it took much more energy to achieve the splitting. At the end of the lecture Rutherford said this:

> While the overall efficiency of the process rises with increase of energy of the bombarding particle, there seems to be little hope of gaining useful energy from atoms by such methods. On the other hand, the recent discovery of the neutron and the proof of its extraordinary effectiveness in producing transformations at very low velocities opens up new possibilities, if only a method could be found of producing slow neutrons in quantity with little expenditure of energy. At the moment, however, the natural radioactive bodies are the only known source for generating energy from atomic nuclei, but this is on far too small a scale to be useful for technical purposes.[11]

Thus, Rutherford was no longer publicly dismissing the possibilities of useful atomic energy. Then, in the Lockyer lecture, certainly thinking of leaders like Hitler, Rutherford made a point of separating men of science from the use of science:

> It is of course true that some of the advances of science may occasionally be used for ignoble ends but this is not the fault of the scientific man, but rather of the community which fails to control this prostitution of science . . . It is sometimes suggested that scientific men should be more active in controlling the wrong use of their discoveries. I am doubtful however whether even the most imaginative scientific man, except in rare cases, is able to foresee the ultimate effect of any discovery.[12]

Rutherford undoubtedly sensed that another world war would unleash the same engines of science and technology that had created new ways and waves of killing in World War I: poison gas, airplanes, and the machine gun. But whatever he was afraid of, he was not alone in downplaying the uses of atomic energy. The other two great scientists of the era, Einstein and Bohr, were doing it as well. Said Bohr, "Not only are such energies, of course, at present far beyond the reach of experiments, but it does not need to be stressed that such effect would scarcely bring us any nearer to the solution of the much-discussed problem of releasing nuclear energy for practical purposes. Indeed the more our knowledge of nuclear reactions advances the remoter this goal seems to become."[13] Einstein put it more colorfully, when he was asked about the chances of releases of massive atomic energy: "It would not be practical. It would be like a blind man in a dark night hunting ducks by firing a shotgun straight up in the air in a country where there are few ducks."[14]

On Thursday, October 13, 1937, Rutherford, still a robust man at sixty-six, was trimming trees in his wife's garden and fell from a low branch. By the next day he was in pain and vomiting. Local doctors diagnosed the problem as an old umbilical hernia that he had aggravated and twisted when he fell. His wife, Mary, an occasional devotee of unconventional medicine, called in a local masseuse who had sometimes been helpful. When the vomiting continued, Dr. Thomas Dunhill, a well-known surgeon, was called down from London to operate on the strangulated hernia. It was a fairly routine operation even in those days. After the surgery Rutherford seemed fine at first, but then he began to fade. Sitting at his bedside, Mrs. Rutherford read and wrote letters, this one to Chadwick in the early hours of October 19:

> Dear Jimmy,
> It's been a long time since I called you that, but one clings so to old friends in time of trouble. My husband is only hanging by a thread. The operation was strangulated hernia with no gangrene which meant they didn't open the gut, but it was paralysed. They treated it and thought it would be all right ... They operated Friday night. Saturday went well. Sunday he started vomiting all day and they realised things had gone wrong or rather that the bowel had not recovered its elasticity. They have tried everything ... He was so splendidly fit up to Wednesday. For some time he had slight bowel weakness and worn a small pad with plaster, but no internal trouble.[15]

Rutherford died later that day. His last words to his wife were, "I want to leave 100 pounds to Nelson College. You can see to it. Remember a hundred to Nelson."[16]

"It was a sunny, tranquil morning, the kind of day on which Cambridge looks so beautiful," said C. P. Snow. "I had just arrived at the crystallographic laboratory, one of the buildings in the old Cavendish muddle . . . Someone put his head around the door and said: 'The Professor's dead.' I don't think anyone said much more. We were stupefied rather than miserable. It did not seem in the nature of things."[17]

At a scientific congress in Italy, Niels Bohr announced what had happened with tears rolling down his face. Lord Rutherford's total estate, it turned out, was almost exactly seven thousand pounds sterling, a bit less than the amount he had received for winning the Nobel Prize in 1908. He had never applied for a patent for any of his discoveries.* 4

Chadwick wrote the obituary that appeared in *Nature*, concluding, "He had, of course, a volcanic energy and an intense enthusiasm—his most obvious characteristic—and an immense capacity for work. A 'clever' man with these advantages can produce notable work, but he would not be a Rutherford. Rutherford had no cleverness—just greatness . . . The world mourns the death of a great scientist, but we have lost our friend, our counsellor, our staff and our leader."[18]

The New York Times said, "He was universally acknowledged as the leading explorer of the vast, infinitely complex universe within the atom, a universe that he was the first to penetrate."[19]

Rutherford's ashes were buried in Westminster Abbey,

* During World War I, the Admiralty had applied for a patent—in the names of Rutherford and W. H. Bragg—on the antisubmarine experiments they had conducted. At Rutherford's request, the patent was voided after the war.

close to the tomb of Sir Isaac Newton. C. P. Snow wrote, "His researches remain the last supreme single-handed achievement in fundamental physics. No one else can ever work there again—in the Cavendish phrase—with sealing wax and string."[20]

Epilogue

For almost seven decades after the Cavendish Laboratory opened its gate on Free School Lane in 1871 and the Germans created their research institutes, physics was largely a European affair. That was why Rutherford left North America in 1908. And for more than twenty of those years, to the outside world, physics was Rutherford and Einstein, Einstein and Rutherford. It is hard to exaggerate the impact of the two men on science in the popular mind. Reporters followed them wherever they appeared; newspapers printed whatever they said as words from on high. "He was as original as Einstein," C. P. Snow wrote of Rutherford, "but unlike Einstein, he did not revolt against formal instruction; he was top in classics as well as everything else . . . If he had pushed on with wireless, incidentally, he couldn't have avoided being rich."[1]

More than sixty years later, in 2005, the American science writer Freeman Dyson, writing in the *New York Review of Books*, said, "The human spirit expresses itself as eloquently in the work of the human hands as in the work of human minds. Rutherford was supreme as an experimenter and Einstein was as a theorist, but each held the other in deep respect. Both of them understood that the human spirit is at its best when hands and minds are working together."[2]

The best expression of what these men did, I think, was in an article in *The New York Times Magazine* on May 24, 1936. The writer was the newspaper's science editor, Waldemar Kaempffert:

Suppose that nobody on earth had ever heard a piece of music. Then suppose that Beethoven's Fifth Symphony is played over and over again by invisible musicians. The physicist's problem is to devise an apparatus which will sift out one note from another and analyze it, infer what kind of invisible instruments produce the sounds, deduce the rules followed in determining what notes should be played and how long and how loudly.

It is not likely that he would succeed in imagining violins and clarinets or even musicians blowing into horns. He would postulate merely vibrating bodies. These would meet his requirements. Even with this simplification the odds against his completely solving the mystery of Beethoven's Fifth Symphony would be heavy.

Solving the problem of the invisible atom, which is far more complex than any orchestra, and which emits light and heat and other forms of energy in more intricate ways than sound is emitted by musical instruments, is infinitely difficult.

Four days after the great Royal Albert Hall rally of 1933, Einstein and his wife sailed to the United States. A position had been created for him at the new Institute for Advanced Studies at Princeton University, after he had rejected British offers arranged by Rutherford. Knowing he could never return to Berlin, Einstein said he preferred America, that England was a country that just had too many butlers.

In Paris, Langevin said it was as if the pope had decided to move from the Vatican to the New World. After the devastation of the European war that would begin in 1939, the center of scientific power and progress inevitably moved away from Europe—away from Germany, of course, but also away from England. One reason was that the vast wealth of America could pay for "Big Science" and finance private companies' efforts to develop more precise and accurate research equipment.

The continental scientists, helped by the Academic Assistance Council, and many of the Englishmen, too, ended up with new lives and careers not in Cambridge or Manchester, but at university towns in the United States. Many, including several "Rutherford men," were absorbed by the Allied atomic bomb project in places like Chicago; Hanford in the state of Washington; Oak Ridge, Tennessee; and Los Alamos, high on a beautiful mesa in New Mexico. 3

In the end more than six hundred thousand people, military and civilian, were involved in what was called the "Manhattan Project." Chadwick, in fact, with Mark Oliphant as one of his assistants, was the director of the British team that came to the United States in September 1943 after the Allies decided to merge their bomb projects—and the Americans realized that their greatest need was experienced experimental physicists. T. E. Allibone was also working at Los Alamos. Most of the Americans working at Los Alamos knew the real identity of "Nicholas Baker," who was listed as a consultant to the British Directorate of Tube Alloys. Those who were not in the know wondered at first why "Baker" was treated with obvious deference. "Mr. Baker" was Niels Bohr. His mother was Jewish, and he had escaped from Denmark with British help only days after the Nazis had decided to include him in a scheduled roundup of "undesirable aliens."

War had taken over science. Otto Hahn, Rutherford's first important student at McGill had discovered, or finally recognized, nuclear fission in 1938. Working secretly with his longtime collaborator Lise Meitner, whose parents were Jewish—who had saved herself by fleeing to Sweden—Hahn had successfully split the nucleus of the heaviest element, uranium-238, into nuclei of barium and krypton, a result so unlikely that other physicists, including Rutherford, had done it without recognizing what had happened. It was Meitner and her nephew, Otto Frisch, studying Hahn's results in exile, who realized the chemistry involved. The rest is history.

A little more than one-half of 1 percent of uranium's weight is uranium-235, an unstable isotope with a nucleus that includes 146 neutrons. It turned out that a single "slow" neutron with its speed reduced and controlled by being shot through "heavy water"—a product of the hydrogen isotope predicted by Rutherford in 1920 but not discovered until 1933—was all it had taken to split a uranium atom. Then it was shown that 2.5 neutrons are released by the one atom of uranium-235 and then speed off in any direction, splitting other uranium atoms in a "chain reaction"—releasing, within one-millionth of a second, the energy that held together those atoms.*

---

* In 1941, both British and American teams discovered another radioactive isotope—plutonium-239—which proved to be more fissionable than

Thus, Rutherford, along with Bohr and Einstein, had been wrong in his first instincts that the bonding energy of atoms could not be controlled. In fact, it could be manipulated or harnessed to produce electric power, and—as Rutherford said in self-contradiction—it was also true that some fool in a laboratory could blow up the world. Now men who were not fools were trying to find out how to do that, trying to build an atomic bomb. The first controlled chain reaction was carried out on December 2, 1942, by Enrico Fermi, who had been a regular Rutherford correspondent and who had decided to flee fascist Italy because his wife was half Jewish.

It was a time of choosing for the physicists touched and inspired by Ernest Rutherford. Otto Frisch, the Austrian who recognized fission, was part of the successful American team at Los Alamos working under J. Robert Oppenheimer, an American theoretical physicist who had studied at Cavendish. Hahn, in Berlin, unable to leave Germany and goaded by Hans Geiger (who was almost certainly a proud Nazi, a member of the National Socialist Party), did enough work on the German bomb project to keep his laboratory open and working. "If my work leads to a nuclear bomb for Hitler," he told Geiger, "I will commit suicide."[3] Meitner, living in exile and poverty in Sweden, was asked by Bohr to work with the Americans. She refused, saying she would never work on bombs.

---

uranium-235. The Americans, led by Fermi, also discovered that graphite could be used as a neutron moderator in place of heavy water. The German atomic bomb project, which included Werner Heisenberg, failed because the carbon it used as a modifier—after Allied air raids destroyed its heavy-water production facilities in occupied Norway—was not as pure and therefore not as effective as graphite.

And what of Kapitsa, who lived until 1984? Many scientists believe that he was critical to the eventual success of the Soviet's atomic and hydrogen bomb projects, though he always denied that he worked on either one. The power of his name and the mystery of his later life was affirmed when the Soviets put the first satellite, Sputnik, into orbit. *The New York Times*, for one, used a photograph of Kapitsa and Reuters dispatch under a 1957 headline:

KAPISTA SKILL PUT SOVIET IN SPACE
Scientist, Once Anti-Red
Gained His Knowledge of Cosmos in Britain[4]

There were many surprises for me along the way in writing this book, but the biggest by far was that Americans who did not work in the sciences did not seem to know who Rutherford was. Although there were new triumphs at Cavendish—particularly the 1959 chemical analysis of deoxyribonucleic acid (DNA) by James Watson and Francis Crick—the decline of British science began or continued after Rutherford's death. Part of that, sadly, was his own enduring legacy of closing laboratories at night. American students at Cavendish in the 1960s remember the heat in their laboratories being turned off at night and on weekends, leaving them to work in overcoats and gloves.

The official history of Cavendish, published in 1974, records, "The decline of the Cavendish Laboratory as the world's leading centre of atomic physics . . . was an inevitable consequence of the war and the transformation of atomic physics to 'Big Science,' to which the tradition of and structure of Cambridge University were ill-adapted."[5] Rutherford's biographer, David Wilson, ends his 639-page book by talking about the influence of the Einstein image of physicists celebrated in the United States. The last sentence reads, "And so it has come about that the American view of the history of nuclear physics has prevailed."[6]

Is that true? Well, the president of the United States, Bill Clinton, celebrating the American century in his 1997 inaugural address, proclaimed, "Along the way, Americans split the atom." No one, to my knowledge, rose to contradict that proof that the winners write the history.

Or perhaps it is too easy now to take for granted the work of genius—as if we always knew the secrets of the atom and there was no more to learn and understand about science. What Ernest Rutherford and his boys wrought with only their minds and their hands changed the world and how we see it—the true test of genius—but it can be bought pretty cheaply today. Catalogs for high school science departments in the United States advertise a modernized version of Rutherford's scattering experiment with sealed apparatus and a digital Geiger counter for $8,475.* Just push a button, and there it is! But one hundred years ago, the mysteries and even the existence of the subatomic world were at least as far away as the edge of the universe is now. Rutherford and men like him changed that, just as other men and women will almost certainly map it all one day. That is what scientists do. That is what they have been doing since the Greeks thought the atom was a hard little fellow—and since a science teacher in New Zealand told that to a boy from a farm on the frontier of South Island.

---

* This price is from the Klinger Educational Products Corporation: "Catalogue of Physics Experiments." The company is headquartered in New York.

Notes

Introduction

1. Brian Sweeney and Jacqueline Owens, "Ernest Rutherford: Atom Man," http://nzedge.com/heroes/rutherford.html (accessed February 2007).

2. "Kapitsa Skill Put Soviet in Space," *New York Times*, November 10, 1957.

3. David Wilson, *Rutherford, Simple Genius* (London: Hodder and Stoughton, 1983), 296. Sir Charles Darwin, who was at lunch with the Rutherfords that December day in 1910, told the story at a Rutherford Jubilee celebration in 1962.

4. Arthur Stewart Eve, *Rutherford; Being the Life and Letters of the Rt Hon. Lord Rutherford, O. M.* (New York: Macmillan, 1939), 101.

5. Frank Corvino, unpublished laboratory notes, November 5, 2005.

6. Wilson, *Rutherford, Simple Genius*, 495.

7. Rutherford to the *London Daily Herald*, quoted by Freeman Dyson, "Seeing the Unseen," *New York Review of Books*, February 24, 2005, 12.

Chapter 1

1. Jeffrey A. Auerbach, *The Great Exhibition of 1851: A Nation on Display* (New Haven, CT: Yale University Press, 1999), 59.

2. Arthur Stewart Eve, *Rutherford; Being the Life and Letters of the Rt Hon. Lord Rutherford, O. M.* (New York: Macmillan, 1939), 11.

3. Oral History Archives of the University of New Zealand.

4. Recounted by Eugene Grayland in his book *Famous New Zealanders* (Christchurch: Whitcombe and Tombs, 1968), as quoted by Brian Sweeney and Jacqueline Owens in "Ernest Rutherford: Atom Man," http://nzedge.com/heroes/rutherford.html (accessed February 2007). The archive of The University of New Zealand maintains a collection of oral histories by men and women who knew the great man, beginning with his mother and sisters. The principal source of Rutherford's early days in England is a collection of Ernest's letters to and from his fiancée, Mary Newton, and his mother, Martha Thompson Rutherford—many of them first published by Arthur Eve (with Lady Rutherford's permission and selection)—which are now in the archives of Cambridge University. And, of course, Rutherford was a great storyteller himself, contributing quips and tales faithfully remembered by students and associates. The mythology that surrounds Rutherford in the land of his birth was emphasized in a letter replying to my inquiry to Gabor Toth of the Wellington City Libraries. Toth added this postscript: "Just a friendly bit of advice should you correspond with any other New Zealand based institutions—it is not necessary to explain who Rutherford was or his achievements to any New Zealander. He is a national hero and was recently ranked as the greatest New Zealander of all time."

5. Balfour Stewart, *Primer of Physics* (London: Macmillan, 1870), 1.

6. Robin McKown, *Giant of the Atom* (New York: Julian Messner, 1962), 26.

7. "Lord Rutherford's Early Years," *Times* (London), October 22, 1937, 18.

8. Eve, *Rutherford*, 13. Thomson's letter of September 24, 1895.

9. Ibid., 15. Letter to his mother, October 3, 1895.

10. Ibid., 19. Letter to Mary Newton, December 8, 1895.

11. Letter to Mary Newton, December 8, 1895.

12. Ibid.

13. David Wilson, *Rutherford, Simple Genius* (London: Hodder and Stoughton, 1983), 69.

14. C. P. Snow, *A Variety of Men* (New York: Scribner, 1967), 5.

15. Wilson, *Rutherford, Simple Genius*, 90.

16. Ibid.

17. Brian Sweeney and Jacqueline Owens, "Ernest Rutherford: Atom Man," http://nzedge.com/heroes/rutherford.html (accessed January 2007).

18. John Dalton, *A New System of Chemical Philosophy, Part I* (London: R. Rickerstoff, 1808).

19. Wilson, *Rutherford, Simple Genius*, 108.

20. Eve, *Rutherford*, 37.

21. Ibid., 42.

22. Ibid., 39.

23. McKown, *Giant of the Atom*, 50.

24. Diana Preston, *Before the Fallout: From Marie Curie to Hiroshima* (New York: Walker, 2005), 26.

Chapter 2

1. Arthur Stewart Eve, *Rutherford; Being the Life and Letters of the Rt Hon. Lord Rutherford, O. M.* (New York: Macmillan, 1939), 55.

2. Ibid., 56. Letter to Mary Newton, August 3, 1898.

3. Ibid., 52. Letter to Mary Newton, June 2, 1898.

4. Ibid., 55. Letter to Mary Newton, August 3, 1898.

5. The current curator of the Room 101 museum is Jean Barrette. I was taken through by his predecessor, Tommy Mark. An impressive virtual tour, including photographs and diagrams of Rutherford's experiments is available at www.physics.mcgill.ca/museum/rutherford_museum.html.

6. E. A. Rutherford, "A Radioactive Substance Emitted from Thorium Compounds," *Philosophical Magazine*, Series 5, No. 49 (January 1900), 1–14.

7. David Wilson, *Rutherford, Simple Genius* (London: Hodder and Stoughton, 1983), 133.

8. "Death of Lord Rutherford," *Times* (London), October 20, 1937, 7.

9. "Death of Lord Rutherford," *Manchester Guardian*, October 20, 1937, 11.

10. E. N. da C. Andrade, *Rutherford and the Nature of the Atom* (Glouces-ter, MA: Smith, 1978), 55. Letter to his mother, January 5, 1902.

11. Frederick Soddy and John Murray, *Science and Life* (London: J. Mur-ray, 1920). The best account of the Rutherford–Soddy relationship, which both scientists wrote about extensively, is Thaddeus J. Trenn's *The Self-Splitting Atom: A History of the Rutherford–Soddy Collabora-tion* (London: Taylor and Francis, 1977).

12. Wilson, *Rutherford, Simple Genius*, 155.

13. Robin McKown, *Giant of the Atom* (New York: Julian Messner, 1962), 58.

14. Barbara Goldsmith, *Obsessive Genius: The Inner World of Marie Curie* (New York: W. W. Norton, 2005), 105.

15. Diana Preston, *Before the Fallout: From Marie Curie to Hiroshima* (New York: Walker, 2005), 31.

16. Ernest Rutherford and Frederick Soddy, "The Cause and Nature of Radioactivity," *Philosophical Magazine*, December 1902: 370–396.

17. Wilson, *Rutherford, Simple Genius*, 207.

18. Eve, *Rutherford*, 63. Letter to Mary Newton, September 25, 1898.

19. McKown, *Giant of the Atom*, 70.

20. Ibid., 82.

21. C. P. Snow, *A Variety of Men* (New York: Scribner, 1967), 7.

22. Letter to Sir William Crookes, April 29, 1902. Cambridge University Library.

23. Andrade, *Rutherford*, 72. Andrade, an important physicist himself, added, "And this was in 1903! Not until 1942 was atomic energy to be released on a large scale in Enrico Fermi's pile at Chicago . . . Such is the foresight of genius."

24. Ibid., 73.

25. Soddy and Murray, *Science and Life*, 118.

26. Mitchel Wilson, "How Nobel Prizewinners Get That Way," *Atlantic Monthly*, December 1969.

27. Wilson, *Rutherford, Simple Genius*, 255; and Eve, *Rutherford*, 93. Rutherford wrote of the evening in 1924.

28. Andrade, *Rutherford*, 75.

29. Eve, *Rutherford*, 107.

30. Ibid.
31. Wilson, *Rutherford, Simple Genius*, 216. Letter from Arthur Schuster to Rutherford, July 7, 1906.

Chapter 3

1. E. N. da C. Andrade, *Rutherford and the Nature of the Atom* (Gloucester, MA: Smith, 1978), 102.
2. Hans Geiger and Ernest Marsden, "The Laws of Deflexion of Alpha Particles through Large Angles," *Philosophical Magazine* 25 (May 4, 1913), 604–623.
3. Chaim Weizmann, *Trial and Error: The Autobiography of Chaim Weizmann* (London: H. Hamilton, 1949), quoted in David Wilson, *Rutherford, Simple Genius* (London: Hodder and Stoughton, 1983), 225.
4. Weizmann, quoted in Wilson, *Rutherford, Simple Genius*, 226.
5. Arthur Stewart Eve, *Rutherford; Being the Life and Letters of the Rt Hon. Lord Rutherford, O. M.* (New York: Macmillan, 1939), 193.
6. W. F. Bynum and Roy Porter, eds., *Oxford Dictionary of Scientific Quotations* (Oxford, England: Oxford University Press, 2004), 145.
7. Eve, *Rutherford*, 352. Address to the Royal Academy of Arts, April 30, 1932.
8. Brian Sweeney and Jacqueline Owens, "Ernest Rutherford: Atom Man," http://nzedge.com/heroes/rutherford.html (accessed May 2007).
9. K. B. Hasselberg, "The Nobel Prize in Chemistry 1908: Presentation Speech," December 10, 1908, http://nobelprize.org/nobel_prizes/chemistry/laureates/1908/press.html (accessed May 2007).
10. Sweeney and Owens, "Ernest Rutherford."
11. Hasselberg, "Nobel Prize."
12. Ernest Rutherford, "The Chemical Nature of the Alpha Particles from Radioactive Substances," Nobel Lecture, December 11, 1908, http://nobelprize.org/nobel_prizes/chemistry/laureates/1908/rutherford-lecture.html (accessed May 2007).
13. Eve, *Rutherford*, 184. Letter to his mother, December 24, 1908.
14. Ibid., 185.
15. Ibid., 186.

Chapter 4

1. David Wilson, *Rutherford, Simple Genius* (London: Hodder and Stoughton, 1983), 291.

2. Hans Geiger, "Memories of Rutherford in Manchester," *Nature* 141 (1938): 244.

3. Wilson, *Rutherford, Simple Genius*, 287.

4. Geiger, "Memories of Rutherford."

5. Lawrence Badash, *Rutherford and Boltwood: Letters on Radioactivity* (New Haven, CT: Yale University Press, 1969), 235.

6. Wilson, *Rutherford, Simple Genius*, 296. Sir Charles Darwin, speaking at the Rutherford Jubilee celebration in 1962.

7. Geiger, "Memories of Rutherford."

8. The paper was published two months later: Ernest Rutherford, "The Scattering of Alpha and Beta Particles by Matter and the Structure of the Atom," *Philosophical Magazine*, Series 6, No. 21 (May 1911), 669.

9. C. P. Snow, *A Variety of Men* (New York: Scribner, 1967), 7.

10. Arthur Stewart Eve, *Rutherford; Being the Life and Letters of the Rt Hon. Lord Rutherford, O. M.* (New York: Macmillan, 1939), 200. Letter from Hantaro Nagaoka, February 22, 1911.

11. Wilson, *Rutherford, Simple Genius*, 241.

12. E. N. da C. Andrade, *Rutherford and the Nature of the Atom* (Gloucester, MA: Smith, 1978), 209.

13. David Sang, *Nuclear and Particle Physics* (Surrey, England: Thomas Nelson, 1995), 578.

14. Diana Preston, *Before the Fallout: From Marie Curie to Hiroshima* (New York: Walker, 2005), 43.

15. Ibid., 45.

16. Wilson, *Rutherford, Simple Genius*, 267.

17. Andrade, *Rutherford*, 124.

18. Wilson, *Rutherford, Simple Genius*, 276.

19. Author's interview with de Gennes, Paris, August 12, 2006.

20. Wilson, *Rutherford, Simple Genius*, 540.

21. Robin McKown, *Giant of the Atom* (New York: Julian Messner, 1962), 112. Rutherford, of course, told the "Sir Ernest" story himself to anyone who would listen.

22. Ibid., 175.

Chapter 5

1. David Wilson, *Rutherford, Simple Genius* (London: Hodder and Stoughton, 1983), 307.

2. Arthur Stewart Eve, *Rutherford; Being the Life and Letters of the Rt Hon. Lord Rutherford, O. M.* (New York: Macmillan, 1939), 233.

3. J. L. Heilbron, *H. G. J. Moseley; the Life and Letters of an English Physicist, 1887–1915* (Berkeley: University of California Press, 1974), 270.

4. Wilson, *Rutherford, Simple Genius*, 270.

5. E. N. da C. Andrade, *Rutherford and the Nature of the Atom* (Gloucester, MA: Smith, 1978), 113.

6. Hans Geiger, "Memories of Rutherford in Manchester," *Nature* 141 (1938): 244.

7. John Masefield, *Gallipoli* (London: Macmillan, 1916), 206.

8. Richard Rhodes, *The Making of the Atomic Bomb* (New York: Simon & Schuster, 1986), 96.

9. Rutherford, "Obituary of H. G. Moseley," *Nature* 96 (September 19, 1915): 33–34.

10. Daniel J. Kevles, *The Physicists: The History of a Scientific Community in Modern America* (New York: Knopf, 1978), 113.

11. Wilson, *Rutherford, Simple Genius*, 405.

12. Robin McKown, *Giant of the Atom* (New York: Julian Messner, 1962), 130.

13. Wilson, *Rutherford, Simple Genius*, 348. The stories of Rutherford and his "boys" in World War I recounted here are taken mostly from Wilson's account. The government records and letters that made his account possible in 1984 were considered "official secrets" for decades before Wilson's research in the 1970s and 1980s. Those wartime records and letters are now available in the Public Records Office in Kew, England, and the archives of Cambridge University. The Cambridge papers include Rutherford's letters and his notebooks for the period, which record the progress he had made on what we now call "sonar." Other important accounts of the war years are contained in the letters of Hans Geiger and in Andrew Brown's 1997 biography of James Chadwick (*The Neutron and the Bomb: A Biography of Sir James Chadwick* [Oxford, England: Oxford University Press]).

14. Ibid., 366.

15. Arthur Stewart Eve, *Rutherford; Being the Life and Letters of the Rt Hon. Lord Rutherford, O. M.* (New York: Macmillan, 1939), 249. Letter to his mother, December 15, 1915.

16. Wilson, *Rutherford, Simple Genius*, 363.

17. Ibid., 379.

18. Ibid., 381.

19. Ibid., 373.

20. Ruth Moore, *Niels Bohr: The Man, His Science & the World They Changed* (New York: Knopf, 1966), 90.

21. Wilson, *Rutherford, Simple Genius*, 394.

22. *New York Times*, January 8, 1922.

23. "Way to Transmute Elements Is Found: Dream of Scientists for a Thousand Years Achieved by Dr. Rutherford," *New York Times*, January 8, 1922, 34.

24. Ibid.

25. Ernest Rutherford, "Atomic Particles and Their Collisions with Light Atoms," *Philosophical Magazine* 37 (1919): 581–587.

26. Wilson, *Rutherford, Simple Genius*, 410.

27. Ibid., 413.

Chapter 6

1. E. N. da C. Andrade, *Rutherford and the Nature of the Atom* (Gloucester, MA: Smith, 1978), 162.

2. Mark Oliphant, *Rutherford: Recollections of the Cambridge Days* (Amsterdam: Elsevier, 1972). Almost without exception, Rutherford's "boys" wrote books or essays about their days with "The Prof." The idiosyncrasies and very British ways of Thomson and Rutherford are described at length by Wilson, and they survived at Cambridge for decades after the deaths of those great men.

3. David Wilson, *Rutherford, Simple Genius* (London: Hodder and Stoughton, 1983), 502.

4. Brian Cathcart, *The Fly in the Cathedral: How a Group of Cambridge Scientists Won the International Race to Split the Atom* (New York: Farrar, Strauss and Giroux, 2004), 114.

5. Robin McKown, *Giant of the Atom* (New York: Julian Messner, 1962), 91.

6. Wilson, *Rutherford, Simple Genius*, 421.

7. Cathcart, *Fly in the Cathedral*, 128.

8. McKown, *Giant of the Atom*, 158.

9. Andrade, *Rutherford*, 169; and Wilson, *Rutherford, Simple Genius*, 444.

10. James Chadwick, "Some Personal Notes on the Search for the Neutron," *Proceedings of the Tenth International Congress of the History of Science [Ithaca, NY, 1962]* (Paris: Hermann, 1964), Vol. 1, 159–162. In his experiments, Chadwick used the radium decay product, radium F, which later became known as polonium-210, and then became a subject of popular interest in late 2006, when a former Soviet police official died in London after persons unknown planted a speck of the toxic isotope in tea that he was drinking at a Japanese restaurant.

11. Andrade, *Rutherford*, 172.

12. Arthur Stewart Eve, *Rutherford; Being the Life and Letters of the Rt Hon. Lord Rutherford, O. M.* (New York: Macmillan, 1939), 291.

13. George Crowe, quoted in Wilson, *Rutherford, Simple Genius*, 569.

14. "Pictures Electrons Speeding in Atom: Sir Ernest Rutherford Says Some Whirl Around at Rate of 93,000 Miles a Second," *New York Times*, September 13, 1923, 3.

15. Ernest Rutherford, "The Annual Address of the President," *Proceedings of the Royal Society* 117 (1927): 300.

16. Ibid.

17. Freeman Dyson, "Seeing the Unseen," *New York Review of Books*, February 24, 2005, 11.

18. *New York Times Book Review*, March 9, 1924.

19. Lawrence Badash, *Kapitza, Rutherford and the Kremlin* (New Haven, CT: Yale University Press, 1985), 5 (footnote).

20. Diana Preston, *Before the Fallout: From Marie Curie to Hiroshima* (New York: Walker, 2005), 69.

21. Ibid., 82.

22. C. P. Snow, *A Variety of Men* (New York: Scribner, 1967), 17.

23. Andrade, *Rutherford*, 189.

24. Andrew Brown, *The Neutron and the Bomb: A Biography of Sir James Chadwick* (Oxford, England: Oxford University Press, 1997), 58.

25. Wilson, *Rutherford, Simple Genius*, 498.
26. Snow, *Variety of Men*, 7.
27. Wilson, *Rutherford, Simple Genius*, 500.
28. Ibid., 499.
29. Ibid., 505.
30. *New Chronicle* (London), April 24, 1935. Soviet Ambassador Ivan Maisky's letter.
31. Wilson, *Rutherford, Simple Genius*, 528.

Chapter 7

1. Brian Cathcart, *The Fly in the Cathedral: How a Group of Cambridge Scientists Won the International Race to Split the Atom* (New York: Farrar, Strauss and Giroux, 2004), flap copy.
2. Author's interview with Daniel Martire, August 2006.
3. "Atom Blasting," *Time Magazine*, March 9, 1931.
4. Arthur Stewart Eve, *Rutherford; Being the Life and Letters of the Rt Hon. Lord Rutherford, O. M.* (New York: Macmillan, 1939), 342.
5. David Wilson, *Rutherford, Simple Genius* (London: Hodder and Stoughton, 1983), 540.
6. C. P. Snow, *A Variety of Men* (New York: Scribner, 1967), 4.
7. E. N. da C. Andrade, *Rutherford and the Nature of the Atom* (Gloucester, MA: Smith, 1978), 175.
8. "To Speed Hydrogen to Break Up Atoms, California Physicists Aim to Hurl Particles at Speed of 37,000 Miles a Second," *New York Times*, September 20, 1930, 5.
9. Cathcart, *Fly in the Cathedral*, 185. Letter from Ernest Walton to Winifred Wilson, February 12, 1930.
10. "A Target of Lithium Bombarded with High Velocity Protons—Experimenters and Apparatus," *Illustrated London News*, June 11, 1932, 70.
11. "Neutron," *Time Magazine*, March 7, 1932.
12. R. van de Graaff, K. T. Compton, and L. C. van Atta, "Electrostatic Production of High Voltage for Nuclear Investigation," *Physical Review* 43 (1933): 149.

13. Merle Tuve, "Biological Effects of Gamma Rays," *Physical Review* 38 (1931): 1919.

14. Ernest O. Lawrence, M. Stanley Livingston, and Milton G. White, "The Disintegration of Lithium by Swiftly-Moving Protons," *Physical Review* 39 (1932): 384.

15. J. D. Cockcroft and E. T. S. Walton, "Letters to the Editor," *Nature* 129 (1932): 242.

16. Ernest O. Lawrence and M. Stanley Livingston, "The Production of High Speed Light Ions without the Use of High Voltages," *Physical Review* 40 (1932): 19–36.

Chapter 8

1. Brian Cathcart, *The Fly in the Cathedral: How a Group of Cambridge Scientists Won the International Race to Split the Atom* (New York: Farrar, Strauss and Giroux, 2004), 224.

2. David Wilson, *Rutherford, Simple Genius* (London: Hodder and Stoughton, 1983), 561. Based on interviews with Walton and Cockcroft; and on Mark Oliphant, *Rutherford: Recollections of the Cambridge Days* (Amsterdam: Elsevier, 1972).

3. Ibid.

4. Cathcart, *Fly in the Cathedral*, 228.

5. Freeman Dyson, "Seeing the Unseen," *New York Review of Books*, February 24, 2005, 12.

6. Cathcart, *Fly in the Cathedral*, 240. Mott went on to become the director of Cavendish from 1954 to 1971.

7. Ibid., 243.

8. J. D. Cockcroft and E. T. S. Walton, "Disintegration of Lithium by Swift Protons," *Nature* 129 (1932), 649.

9. C. P. Snow, "Rutherford and the Cavendish," in *The Baldwin Age*, ed. John Raymond (London: Eyre and Spottiswoode, 1960).

10. *Reynold's Illustrated News*, No. 4 (May 1, 1932), 261.

11. Cathcart, *Fly in the Cathedral*, 248.

12. *New York Times*, May 2, 1932, 1.

13. Cathcart, *Fly in the Cathedral*, 254.

14. Interview with Merle Tuve. Center for History of Physics, American Physics Institute, January 1982.

15. Walton's letter to Winifred Wilson, August 5, 1931.

16. Ibid.

17. Wilson, *Rutherford, Simple Genius*, 430.

18. Ibid., 428.

19. Harold Nicolson, *Public Faces* (New York: Popular Library, 1960), 17.

20. Wilson, *Rutherford, Simple Genius*, 484.

21. Ibid., 543.

22. Mark Oliphant, *Rutherford: Recollections of the Cambridge Days* (Amsterdam: Elsevier, 1972), 60. Letter from Max Born to James Chadwick, August 11, 1954.

23. Arthur Stewart Eve, *Rutherford; Being the Life and Letters of the Rt Hon. Lord Rutherford, O. M.* (New York: Macmillan, 1939), 380. Letter from Johannes Stark to Rutherford, February 28, 1934.

24. Ibid.

25. "Dr. Einstein on Liberty. Albert Hall Speech," *Times* (London), October 4, 1933.

26. "Sir A. Chamberlain on the Nazi Persecution," *Manchester Guardian*, October 4, 1933.

27. Wilson, *Rutherford, Simple Genius*, 488.

Chapter 9

1. Samuel Devons, "Rutherford's Laboratory," in *A Hundred Years and More of Cambridge Physics*, ed. Dennis Moralee (Cambridge, England: Cambridge University Physics Society, 1974), 24.

2. Ibid., 29.

3. Mark Oliphant, *Rutherford: Recollections of the Cambridge Days* (Amsterdam: Elsevier, 1972).

4. Diana Preston, *Before the Fallout: From Marie Curie to Hiroshima* (New York: Walker, 2005), 97.

5. David Wilson, *Rutherford, Simple Genius* (London: Hodder and Stoughton, 1983), 587.

6. Ibid.

7. C. P. Snow, *A Variety of Men* (New York: Scribner, 1967), 9.

8. "The British Association: Breaking Down the Atom: Transformation of the Elements," *Times* (London), September 12, 1933, 6.

9. Wilson, *Rutherford, Simple Genius*, 573.

10. Ibid., 493.

11. E. N. da C. Andrade, *Rutherford and the Nature of the Atom* (Gloucester, MA: Smith, 1978), 210.

12. Wilson, *Rutherford, Simple Genius*, 584.

13. Moore, *Niels Bohr*, 213.

14. Ibid., 213.

15. Andrew Brown, *The Neutron and the Bomb: A Biography of Sir James Chadwick* (Oxford, England: Oxford University Press, 1997), 158.

16. Arthur Stewart Eve, *Rutherford; Being the Life and Letters of the Rt Hon. Lord Rutherford, O. M.* (New York: Macmillan, 1939), 425.

17. Snow, *Variety of Men*, 20.

18. James Chadwick, "Lord Rutherford, Obituary," *Nature* 140 (1937): 749.

19. "Lord Rutherford, Physicist, Is Dead: British Nobel Winner, 66, Famous as Atom Smasher," *New York Times*, October 20, 1937, 1.

20. Snow, *Variety of Men*, 10.

Epilogue

1. "Lord Rutherford, Physicist, Is Dead: British Nobel Winner, 66, Famous as Atom Smasher," *New York Times*, October 20, 1937, 1.

2. Freeman Dyson, "Seeing the Unseen," *New York Review of Books*, February 24, 2005, 12.

3. Richard Rhodes, *The Making of the Atomic Bomb* (New York: Simon & Schuster, 1986), 262.

4. "Kapitsa Skill Put Soviet in Space," *New York Times*, November 10, 1957.

5. Dennis Moralee, ed., *A Hundred Years and More of Cambridge Physics* (Cambridge, England: Cambridge University Physics Society, 1974), 31.

6. David Wilson, *Rutherford, Simple Genius* (London: Hodder and Stoughton, 1983), 639.

BIBLIOGRAPHY

There have been at least a dozen of biographies of Ernest Rutherford, including the official biography written by one of his former students and associates, Arthur Stewart Eve: *Rutherford*. That 451-page volume, published by Cambridge University Press in 1939, includes an invaluable collection of Rutherford's letters to his wife, his mother, and scientists around the world; and it was, in fact, produced under the personal supervision of Lady Rutherford, who controlled the content. The most comprehensive of the biographies is the 639-page *Rutherford, Simple Genius*, by David Wilson, published by Hodder and Stoughton of London in 1983.

I also used material from several shorter books: *Rutherford and the Nature of the Atom*, by Edward Neville da Costa Andrade, a Rutherford colleague at the University of Manchester, published by Peter Smith of Gloucester, Massachusetts, in 1978; *Rutherford and Boltwood: Letters on Radioactivity*, edited by Lawrence Badash and published by Yale University Press of New Haven, Connecticut, in 1969; *Giant of the Atom*, by Robin

McKown, published by Julian Messner of New York in 1962; *The Neutron and the Bomb: A Biography of Sir James Chadwick*, by Andrew Brown, published by Oxford University Press of Oxford, England, in 1997; *Rutherford and Physics at the Turn of the Century*, edited by Mario Bunge and William R. Shea, and published by Dawson and Science History Publications, New York, in 1979 (a collection of papers by physicists, including Lawrence Badash and Norman Feather, issued and discussed at a conference in 1977 on the fortieth anniversary of Rutherford's death).

In addition, I made use of several well-done books covering various periods and advances in the knowledge of physics in which Rutherford was a key player: *The Making of the Atomic Bomb*, a magnificent work of history by Richard Rhodes, published by Simon & Schuster of New York in 1986; *The Fly in the Cathedral: How a Group of Cambridge Scientists Won the International Race to Split the Atom*, by Brian Cathcart, published by Farrar, Strauss and Giroux in New York in 2004; *Before the Fallout: From Marie Curie to Hiroshima*, by Diana Preston, published by Walker Publishing of New York in 2005; *Kapitza, Rutherford and the Kremlin*, by Lawrence Badash, published by Yale University Press in New Haven, Connecticut, in 1985.

I gained a good deal of insight into Rutherford's world through the work of C. P. Snow, who had known Rutherford at Cambridge, particularly *A Variety of Men*, a series of essays published by Charles Scribner's Sons in New York, in 1967; and *The New Men*, published by the House of Stratus in York, England, in 1954 (the sixth of the nine novels that make up Snow's "Strangers and Brothers" series). Ruth Moore's biography, *Niels Bohr*, published by Alfred A. Knopf, New York, in 1966, was also helpful.

I also worked with several standard modern physics textbooks, particularly *Nuclear and Particle Physics*, by David Sang, published in 1995 by the University of Bath in England and Thomas Nelson and Sons of Walton on Thames, Surrey, England.

Four privately published museum guides were valuable: *The Rutherford Museum of McGill University*, written in 1996; and three small volumes published by Cambridge University—*A Hundred Years and More of Cambridge Physics* (1998); *The Cavendish Laboratory: An Outline Guide to the Museum* (1980); and *Selected Apparatus in the Cavendish Museum* (1998).

The most important journalistic archives used in my work were those of *The New York Times* and *Time* magazine in the United States, and the *Times* of London and the *Manchester Guardian* in England. Ernest Rutherford was probably the most famous experimental scientist of the first third of the twentieth century, and his deeds and words were covered extensively almost everywhere in the world. Citations from those journals are generally identified by date in the text of this volume.

Finally, I consulted dozens of physics and university Web sites for details on experiments or dates.

ACKNOWLEDGMENTS

I am grateful to James Atlas of Atlas Books for thinking of me as an author who might contribute to the "Great Discoveries" series published with W. W. Norton & Company, and for his interest and support in the idea of re-creating Ernest Rutherford's 1909 "scattering" experiment, which I did at Stevens Institute of Technology in Hoboken, New Jersey. That experiment and my scientific reeducation would have been impossible without the skill and friendship of Dr. Kurt Becker, professor of physics and former head of the Physics Department, and the lab team he put together for me there: George Wohlrab, Frank Corvino, and Damien Marianucci. (After we completed our work, Becker was named Associate Provost for Technology at the Polytechnic University of New York.)

I also owe a great deal to two of my classmates at Stevens: Dr. Donald Merino, professor of industrial engineering at Stevens, who helped me put the project together; and Dr. Daniel Martire, retired chairman of the Department of Chemical Engineering at Georgetown University, who tried to keep me from embarrassing myself, just as he did when we lived across the hall from each other at school. I also want to thank all the professors and curators, particularly Dr. Tommy Mark at McGill and Dr. Peter Bystricky at Cambridge, who educated me about Lord Rutherford's methods and equipments in the laboratories and museums of McGill University in Montreal, the University of Manchester, and the Cavendish Laboratory at Cambridge University. I learned a good deal, too, from another writer in this series, Barbara Goldsmith, author of *Obsessive Genius: The Inner World of Marie Curie*.

I am also indebted to Jessica Fjeld of Atlas Books, who handled the business of editing the manuscript, along with John Oakes, Stephanie Hiebert, and Janet Lee. In London, Vivien Burgess did valuable research for me in the archives of the *Times* of London and the *Manchester Guardian*. Sue Gifford managed all that and more for two years.

My agent, Amanda Urban of International Creative Management, handled the financial arrangements and the other details that make life possible for an author. Alice Mayhew, my longtime editor at Simon & Schuster, was kind enough to allow me to work with her competitors and friends at Atlas Books and W. W. Norton. But, in many ways, this book was hardly about the money. When I told Kurt Becker that I thought we should have a letter of agreement about paying for the significant costs of doing the experiment, he said, "Forget that. This is for fun. We're all going to have fun doing this."

As always, my wife, Catherine O'Neill, and our children put up with that faraway look and distant peace writers always seem to have when they are working. They gave me a good deal of room on this one because they saw how much fun I was having.

Index

Richard Reeves, senior lecturer at the Annenberg School of Communication at the University of Southern California, has written a dozen books on American and world politics, including a trilogy on the modern American presidency: *President Kennedy: Profile of Power* (1993), *President Nixon: Alone in the White House* (2001), and *President Reagan: The Triumph of Imagination* (2006). The three were published by Simon & Schuster in the United States. Reeves is a syndicated columnist and former chief political correspondent of *The New York Times*. He graduated from Stevens Institute of Technology in Hoboken, New Jersey, with a degree in mechanical engineering and worked as an engineer before becoming a journalist.